The ESL SCIENCE Book

The ESL SCIENCE Book

John F. Chabot
Kirk Schriefer

Illustrations: Rosalind Anderson & Cynthia DiSimone & Jeannette Julich

FULL BLAST PRODUCTIONS

IN CANADA

FB Productions
70 Allan Drive
Saint Catharines, Ontario
L2N 1E9

IN THE UNITED STATES

FB Productions
1623 Military Road #370
Niagara Falls, New York
14304-1745

Phone: 905-937-1569
Fax: 905-937-2657
E-mail: fbp@cogeco.ca
Website: www.fullblastproductions.com

Library and Archives Canada Cataloguing in Publication

The ESL science book / John F. Chabot ... [et al.] ; illustrations: Rosalind Anderson & Cynthia DiSimone & Jeannette Julich.

ISBN 1-895451-62-0

1. English language--Textbooks for second language learners. 2. Science--Textbooks. I. Chabot, John F., 1959- II. Anderson, Rosalind III. DiSimone, Cynthia, 1959- IV. Julich, Jeannette

PE1128.E842 2006 428.2'4 C2006-904356-6

Printed in Canada

INTRODUCTION

The ESL SCIENCE Book is a reproducible ESL/EFL reading and discussion text for intermediate or higher level students. It offers a number of important design features that make both teaching and learning easy and enjoyable.

1) **Contextualized learning:** Each reading selection focuses on a single topic, and is followed by vocabulary and comprehension exercises that review and expand upon the theme of the reading.

2) **Relevant content:** Each unit examines a topic that is interesting and informative.

3) **Lively journalistic style:** Although the passages are written for English language learners, their style remains vivid and authentic.

4) **Variety of exercises:** The numerous exercises offer a range of integrated activities for each unit -- getting the main idea, basic comprehension, finding details, inferences, interpretation and extension of important concepts, vocabulary study, a word puzzle, and a cloze exercise.

5) **Illustrations:** There is a clear and evocative illustration at the beginning of each unit, which can be used for pre-reading exercises such as activating prior knowledge and vocabulary brainstorming, or as a basis for discussion.

6) **Answer Key:** The text includes a full answer key for every closed-ended question in every unit.

7) **Reproducible Format:** Purchase of an original copy of this text gives the purchaser explicit permission to reproduce pages at will.

This package can be used quite simply as it is presented in these pages. In fact, it was carefully planned to be effective in that way. However, resourceful teachers will most likely want to consider one or more of the following suggestions:

- **Pre-Activity:** Instead of immediately beginning to read the passage, have students start with a discussion or game to encourage them to think about the theme of the unit. The picture or title at the beginning of the unit can be useful in this connection; so can newspaper or magazine illustrations or headlines, or current television or radio news items.

- **Order of Exercises:** By all means, use all the activities provided for each unit, and allow sufficient time for these to be completed thoroughly (including time for small groups to discuss alternate answers and so on). But build in variety by changing the order in which the exercises are done, and by varying the designation of activities for in-class or at-home assignments.

- **Intensity of Work:** Be sure to take full advantage of the wide range of different kinds of involvement that these materials offer. At one end of the continuum, for instance, the *Interpretation* questions can be used to inspire free, creative discussion of themes, values and general ideas. By contrast, at the other end of the continuum, the *Word Power* exercises provide an excellent foundation for practice in all the detailed and demanding but very important strategies for independent vocabulary development: not just careful contextual reading and word-part analysis, but also dictionary and thesaurus use. Resourceful attention to this kind of varied involvement with the materials will make teaching and learning more enjoyable and more effective, too.

- **Post-Activity:** Once the class has finished the set of exercises in the text itself, think about rounding off the cycle with a post-activity that clearly links the book and the schoolroom to the wider world. Learners might pursue the theme or content of a given unit by making a bulletin-board collage of their own art work and/or comments, writing letters to a relevant person or institution, going on a visit, watching a video...you will find many ways to reinforce the vocabulary, grammar and content one more time while you also build the self-image of learners as competent language-users for real-life purposes.

- **Flexibility:** There is no problem with using these units in alphabetical order, just as provided, but in many circumstances their best application will be as a flexible database of content based readings that you can access, to tap into the energy of learners' natural curiosity about such passing interests as news items, current movies, new music, or extra-curricular activities. Dip into this treasure-trove of self-contained units to find the exact passage to exploit a "hot topic." Each term you'll have different students with different experiences, and so your use of the text will differ as well. With this flexible resource, you can modify your teaching to meet your students' needs!

TABLE OF CONTENTS

Dinosaurs

1 The word dinosaur is Greek for "terrible lizard" and for many people the term suggests huge creatures with horns, sharp teeth and long tails. But believe it or not, some of the first dinosaurs, called thecodonts, were no bigger than turkeys. Thecodonts lived over 200 million years ago.

2 It took millions of years but dinosaurs eventually grew larger and divided into two categories: carnivores, or meat-eaters, and herbivores, plant-eaters. The plant-eaters lived near warm swampy areas that provided lush vegetation and an abundant supply of plants to eat. The Apatosaurus species was 30 feet (9 meters) tall and 90 feet (27.5 meters) long from head to tail.

3 The largest meat-eating dinosaur was Tyrannosaurus Rex. At 45 feet (14 meters) long and 20 feet (6 meters) tall, with eight-inch dagger-like teeth, the Tyrannosaurus was the worst enemy of the plant-eating dinosaurs. With its long, powerful hind legs and sharp, curved claws, the T-Rex would hide near a swamp and attack any plant-eating dinosaur that came near the water to drink.

It is not known how all dinosaurs reproduced, though it is generally accepted that some dinosaurs gave birth by laying eggs. Fossil eggs and egg fragments have been found in quite a few places around the world. 4

Dinosaurs prospered and ruled the earth for 150 million years. Scientists believe that over millions of years their 5 numbers declined until about 65 million years ago when they became extinct. For a long time the accepted theory was that a slow change in climate led to the disappearance of many plants. Without plants to eat, the plant-eating dinosaurs died and then the meat-eating dinosaurs died when their source of food was gone.

6 But a new extinction theory was proposed in the late 1970s. An American scientist, Luis Alvarez, detected the chemical iridium in ancient soil from the end of the Age of Dinosaurs. The iridium was found in soil samples from around the world. Usually, this rare chemical is found only deep in the earth and in meteorites that fall from space.

7 Alvarez reasoned that the iridium must have been spread across the planet's surface when a giant comet struck the earth. This collision would have caused thick dust to be thrown into the atmosphere, which would have blocked out the sun. Such an environmental disaster would have resulted in a long period of darkness and cold. Alvarez suggests that such a change in their world caused the dinosaurs to die abruptly.

Dinosaur fossils have been discovered worldwide, but most of our knowledge about dinosaurs comes from the fossils found in North America from 1870 to the early 1900s. The greatest discoveries came from a layer of rocks called the Morrison Formation, which runs through six American states from Montana to Oklahoma. And in the Alberta badlands of Canada, several expeditions between 1910 and 1917 discovered fossils of four new species of dinosaurs. 8

The passing of sixty-five million years has made it impossible for Alvarez or any other scientist to positively prove why the dinosaurs became extinct. In fact, we may never know for sure how the dinosaurs died. But surely the amazing thing is that they lived at all! 9

THE MAIN IDEA

Circle the letter of the sentence or phrase which best describes the main idea of the article about dinosaurs. Be prepared to support your answer.

a) The biology of dinosaurs.
b) How dinosaurs ruled the earth.
c) Discoveries of dinosaur fossils.
d) The rise and fall of dinosaurs.

UNDERSTANDING WHAT YOU READ

If you can, answer these questions from memory. If you cannot, look back at the article.

1) What does the word dinosaur mean?

____

2) What two basic kinds of dinosaurs lived on the earth?

____

3) Give the size of one large and one small dinosaur.

____

4) When did dinosaurs become extinct?

____

REMEMBERING DETAILS

Write TRUE or FALSE under each statement. If the statement is false, write the statement correctly.

1) Tyrannosaurus Rex was the largest meat-eating dinosaur.

____

2) One thing we know for sure about dinosaurs is that they all laid eggs.

____

3) Alvarez developed the theory that the dinosaurs died because of a slow change in the climate.

____

4) Scientists have found dinosaur fossils only in North America.

____

INFERENCES

Based on the article, circle the letter of the best sentence completion.

1) If the change in climate theory is correct, the extinction of the dinosaurs suggests that they were...

a) not strong enough.
b) not adaptable enough.
c) not intelligent enough.
d) not big enough.

2) Dinosaurs died out millions of years ago,...

a) but we have a very clear idea about exactly how they lived.
b) and so they do not receive much attention now.
c) but people are still very interested in them.
d) and yet the effects of their extinction are still being felt.

INTERPRETATION

1) If dinosaurs had not died out, how do you think humans and dinosaurs would have interacted?

2) Many children and adults are fascinated by dinosaurs, and there have been many books and movies about them. Do you share this fascination? Why or why not?

3) Do you think it is important to spend time and money on research about dinosaurs? Why or why not?

WORD POWER

Circle the letter of the word that means the same as the word on the left.

1) swampy	a) wet	b) strongly	c) vertically
2) lush	a) dense	b) sparse	c) colorful
3) dagger-like	a) like fingers	b) like teeth	c) like knives
4) hind	a) thick	b) back	c) long
5) reproduced	a) grew	b) bred	c) manufactured
6) abruptly	a) suddenly	b) mysteriously	c) completely

CROSSWORD PUZZLE

ACROSS:

1) Such an ____________________ disaster would have resulted in a period of darkness and cold.

4) The term suggests huge ____________________ with horns, sharp teeth and long tails.

6) Iridium is found only deep in the earth and in ____________________ that fall from space.

11) Dinosaurs ____________________ and ruled the earth for 150 million years.

13) The collision would have caused thick ____________________ to be thrown into the atmosphere.

14) ____________________ eggs have been found in quite a few places around the world.

DOWN:

2) The ____________________ was found in soil samples from around the world.

3) Many discoveries came from a ____________________ of rocks called the Morrison Formation.

5) The T-Rex would hide near a ____________________ and attack plant-eating dinosaurs.

7) ____________________ were no bigger than turkeys.

8) A new ____________________ theory was proposed in the late 1970s.

9) The Apatosaurus ____________________ was the largest plant-eating dinosaur.

10) Dinosaur fossils have been discovered ____________________.

12) Without plants to eat, the plant-eating ____________________ died.

ANSWER KEY

THE MAIN IDEA

d) The rise and fall of dinosaurs.

UNDERSTANDING WHAT YOU READ

1) The word dinosaur means "terrible lizard."
2) The two basic kinds of dinosaurs were carnivores, which ate meat, and herbivores, which ate plants.
3) Thecodonts were small dinosaurs, only about the size of a turkey; the Apatosaurus was a big dinosaur, about 30 feet tall and 90 feet long.
4) Dinosaurs became extinct about 65 million years ago.

REMEMBERING DETAILS

1) T
2) F Some dinosaurs did lay eggs, but we do not know how all dinosaurs reproduced.
3) F Alvarez's theory is that dinosaurs died as a result of a rapid change in climate, caused when a giant comet hit the earth.
4) F Most of our information about dinosaurs does come from fossils from North America, but in fact dinosaur fossils have been found all over the world.

INFERENCES

1) b
2) c

WORD POWER

1) a
2) a
3) c
4) b
5) b
6) a

CROSSWORD PUZZLE

ACROSS: 1) environmental 4) creatures 6) meteorites 11) prospered 13) dust 14) fossil

DOWN: 2) iridium 3) layer 5) swamp 7) thecodonts 8) extinction 9) species 10) worldwide 12) dinosaurs

Global Warming

1 Since the 1980s global temperatures have been rising at what many scientists feel are alarming rates. If temperatures continue to rise as they have been, scientists predict several types of environmental disturbances. The most commonly known theory of global warming is the greenhouse effect.

2 Greenhouses are made of glass; this lets in the sunlight all the time. The sunlight warms up all the plants and equipment in the greenhouse. The plants and other objects then give off heat but the glass walls of the greenhouse do not let the heat escape--it stays inside. Without proper ventilation the temperature in the greenhouse would soar to unbearable heights and kill all the plants. Another example of this effect is when a car is parked in the sunlight with all its windows rolled up.

The earth can sustain life because the 3
planet's atmosphere works like a well-ventilated greenhouse. In fact, the average temperature on the planet is 60 degrees Fahrenheit (16 degrees Celsius). The problem is that the ventilation system that keeps the planet at the right temperature is getting clogged. Carbon dioxide produced by industry, agriculture and transportation are trapping more and more of the sun's heat inside the atmosphere.

Like a closed-up greenhouse or a 4
locked-up car, the thickening, harmful gases on the earth's lower atmosphere act in the same way. This has resulted

in a slow but steady rise in global temperatures, which some scientists predict will cause the global average temperature to rise as much as 9 degrees Fahrenheit (5 degrees Celsius) before the year 2100.

5 If the carbon dioxide cover around the earth continues to build up, the earth could someday be similar to the planet Venus. Venus is a planet (otherwise very similar to ours) that has a highly advanced greenhouse effect. It has a thick carbon dioxide atmospheric cover that traps the sun's heat and makes the surface of Venus hot enough to melt lead. But the earth is a long way from that kind of scenario right now.

6 Still, the planet faces risks from rising global temperatures that could be catastrophic. Droughts, more frequent tropical storms and extinction of some species could result from global warming. This would seriously disrupt our planet's food production. And hotter weather in the northern and southern arctic regions could result in the melting of polar ice caps, which would cause coastal flooding in ocean-bordering countries.

7 Some scientists are hopeful that nature itself will help reduce the greenhouse effect. One way nature could accomplish this would be by increased cloud cover. More clouds would absorb more heat and therefore lower temperatures. Another thing in nature that could help is plankton, which is a living organism found in oceans. Plankton naturally absorbs carbon dioxide, thus reducing the greenhouse effect.

8 In 1990 a panel of 170 scientists from around the world released a report warning that the greenhouse effect would continue to increase noticeably unless something was done. In 2005, the Kyoto Protocol, which is an amendment to a clean air treaty signed in 1992, came into effect. Over 160 nations have signed and agreed to reduce air pollution.

THE MAIN IDEA

Circle the letter of the sentence or phrase which best describes the main idea of the article about global warming. Be prepared to support your answer.

a) Causes and dangers of global warming.
b) The future of global warming.
c) What is meant by the greenhouse effect.
d) How scientists study global warming.

UNDERSTANDING WHAT YOU READ

If you can, answer these questions from memory. If you cannot, look back at the article.

1) What do scientists predict could be the result if temperatures continue to rise?

---

2) How does the greenhouse effect work?

---

3) How hot is the surface of Venus?

---

4) What was the conclusion of the report released by a panel of scientists in 1990?

REMEMBERING DETAILS

Write TRUE or FALSE under each statement. If the statement is false, write the statement correctly.

1) Since the 1980s scientists have been alarmed by rising global temperatures.

---

2) You could say that the earth is exactly like the car parked in the sun, with all the windows rolled up.

---

3) Before the year 2100, temperatures on earth are predicted to be up to 20 degrees Fahrenheit (about 10 degrees Celsius) higher than now.

---

4) Increased cloud cover is part of the greenhouse effect, making the temperature on earth rise.

INFERENCES

Based on the article, circle the letter of the best sentence completion.

1) Industry, agriculture and transportation...

a) are largely responsible for plankton levels in our oceans.
b) are largely responsible for the studies done on global warming.
c) have responded favorably to new pollution laws.
d) are at least partly responsible for rising temperatures.

2) The example of Venus is important because...

a) it shows what is happening on earth right now.
b) it shows what has happened on earth in the past.
c) it shows what will happen to the earth in future.
d) it shows what might happen to the earth in the future.

INTERPRETATION

1) Is the greenhouse effect just the responsibility of industrialized nations or of big factory owners, or is it everyone's duty to help stop this process?

2) What do you think individual people can do to avoid contributing to global warming? Are you doing those things yourself? Why or why not?

3) Have you already experienced any evidence of global warming? What have you seen? How do these changes affect your life?

WORD POWER

Circle the letter of the word that means the same as the word on the left.

1) alarming	a) alerting	b) troubling	c) boring
2) rise	a) increase	b) decrease	c) stability
3) soar	a) come down	b) go up	c) change around
4) sustain	a) support	b) restrain	c) observe
5) disrupt	a) interfere with	b) go along with	c) put up with
6) restrictive	a) striking	b) limiting	c) lax

CROSSWORD PUZZLE

ACROSS:

2) ____________________ temperatures have been rising at alarming rates.
4) Hotter weather could result in the melting of ____________________ ice caps.
5) More clouds would ____________________ more heat and therefore lower temperatures.
6) The planet's ventilation system is getting ____________________.
8) The average ____________________ on the planet is 60 degrees Fahrenheit (16 degrees Celsius).
10) Scientists are hopeful that ____________________ itself will help reduce the greenhouse effect.
12) A thing that could help is ____________________, which is a living organism found in oceans.

DOWN:

1) Plankton naturally absorbs carbon ____________________.
2) Greenhouses are made of ____________________; this lets in the sunlight all the time.
3) Nature could accomplish this by having an increased ____________________ cover.
6) Rising global temperatures could be ____________________.
7) The most commonly known theory of global warming is the ____________________ effect.
9) Over 160 nations have signed the Kyoto Protocol and agreed to reduce air ____________________.
11) ____________________ is a planet that has a highly advanced greenhouse effect.

ANSWER KEY

THE MAIN IDEA

a) Causes and dangers of global warming.

UNDERSTANDING WHAT YOU READ

1) If temperatures continue to rise scientists predict several types of environmental disturbances.
2) Changes in the earth's atmosphere trap more heat in, making the earth warmer, just as the glass of a greenhouse keeps the heat in. Just like a greenhouse, the earth needs a good ventilation system to prevent it from becoming too hot. Unfortunately the planet's ventilation system is becoming clogged.
3) The surface of Venus is hot enough to melt lead.
4) The report concluded that the greenhouse effect would increase noticeably unless something was done.

REMEMBERING DETAILS

1) T
2) F This statement would be an exaggeration because so far the situation is not that bad. Unlike the locked car, the earth does not yet collect all the heat that reaches it from the sun, and there is still a chance to prevent the worst results.
3) F The prediction is that by then temperatures may be as much as 9 degrees Fahrenheit (5 degrees Celsius) higher.
4) F Increased cloud cover may turn out to cause a reduction in the greenhouse effect by absorbing some of the heat from the sun and therefore stopping it from reaching the earth.

INFERENCES

1) d
2) d

WORD POWER

1) b
2) a
3) b
4) a
5) a
6) b

CROSSWORD PUZZLE

ACROSS: 2) global 4) polar 5) absorb 6) clogged 8) temperature 10) nature 12) plankton

DOWN: 1) dioxide 2) glass 3) cloud 6) catastrophic 7) greenhouse 9) pollution 11) Venus

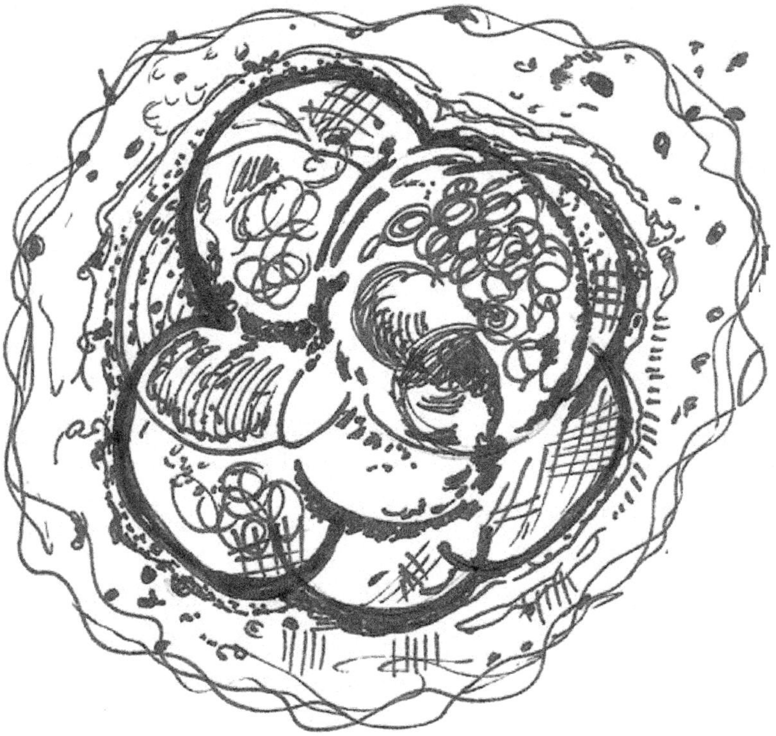

Stem Cells

1 Wouldn't it be great if doctors could replace worn out cells in your body with new ones? Imagine you had problems with your heart or your brain or your spinal chord. What if scientists could grow tissue to specifically fill your needs? Need new heart tissue? The scientists grow it. Have a brain illness like Parkinson's disease? Scientists grow replacement brain tissue for you. This is the promise of stem cells and research into using stem cells to cure damaged or diseased organs and tissues.

2 What is a stem cell? The simplest way to put it is a stem cell is a human cell that has yet to make up its mind to which part of the body it belongs. The cells have yet to become specialized. Through stem cell research, scientists are trying to discover ways to make these uncommitted cells behave the way they want. They want to be able to grow the cells into whatever type of replacement tissue is needed.

3 At present, there are three kinds of stem cell research. The three are embryonic stem cell research, adult stem cell research and research into stem cell cloning.

4 Many stem cells can be found in a human embryo in its early stages of development. In the United States in 1998 two groups of scientists managed to get stem cells from human embryos to survive then reproduce in a laboratory. These scientists have had some success getting stem cells to grow. At present, there is no way for the scientists to control into what the stem cell grows,

but they believe in the future they will find a way. The time may not be too far distant that stem cells are taken from a baby's umbilical cord and kept to use to treat illnesses the baby may have later in its life.

5 Surprisingly, stem cells do exist in adults. Most of the cells in an adult are already part of an organ or a nerve or some other part of the body, but a few uncommitted stem cells are present...very few. In fact, in an adult, stem cells are difficult for scientists to find because there are such a small number of them. Another dilemma for scientists is that adult stem cells are difficult to grow. As with stem cell research with embryos, scientists still need to find the answer how to control what an adult stem cell grows into. Researchers hope that someday stem cells can be taken directly from a sick adult person and grown into the tissue that person needs to help them recover.

6 Stem cell cloning involves taking a cell from the person being treated and cloning a human embryo with the same genetic information as the patient. The embryo is a clone of the person. However, the embryo will not live long. It will have its stem cells harvested and be destroyed. In this case, scientists are only interested in growing stem cells. Like with embryonic and adult stem cell research, scientists are still looking for answers how to control the stem cells. If and when scientists find the answer to this problem, the tissue will be a perfect match with the DNA of the patient.

Many people are opposed to stem cell 7
research. Their opposition can be on moral, ethical or religious grounds. Some people are opposed to the cloning involved in stem cell cloning, and what about killing the cloned embryos? And what about the embryos used in embryonic stem cell research?

Of course the other side of the argument 8
is that a breakthrough using stem cells could change medicine forever and offer humans cures for many of the physical ills that strike us.

THE MAIN IDEA

Circle the letter of the sentence or phrase which best describes the main idea of the article about stem cells. Be prepared to support your answer.

a) A cure for the physical ills that strike humans.
b) Moral issues concerning stem cell research.
c) An explanation of stem cells and stem cell research.
d) How stem cells are grown.

UNDERSTANDING WHAT YOU READ

If you can, answer these questions from memory. If you cannot, look back at the article.

1) What is the promise of stem cell research?

---

2) What is a stem cell?

---

3) Why are there so few stem cells in an adult?

---

4) On what grounds are some people opposed to using stem cells?

REMEMBERING DETAILS

Write TRUE or FALSE under each statement. If the statement is false, write the statement correctly.

1) There are more stem cells in adults than there are in embryos.

---

2) The time may not be too far distant that stem cells are taken from a baby's umbilical cord and kept to use later.

---

3) Scientists know how to control stem cells they get through cloning.

---

4) A breakthrough using stem cells would have no impact on medicine.

INFERENCES

Based on the article, circle the letter of the best sentence completion.

1) The biggest question regarding stem cell research seems to be,...

a) are scientists going to be able to someday control what the stem cells do?
b) are people opposed to stem cell research going to change their point of view?
c) are scientists going to discover something more important than stem cell research?
d) is stem cell research really going to be able to help people with Parkinson's disease?

2) Stem cell research...

a) may someday be an everyday part of medicine.
b) will cure everything.
c) is better for children.
d) will work within ten years.

INTERPRETATION

1) What is your position on stem cell research? Discuss.

2) How important do you think it would be for scientists to discover how to control stem cells? How important for medicine? How important for people?

3) List ten discoveries that have had a major impact on people. Share your list with your classmates.

WORD POWER

Circle the letter of the word that means the same as the word on the left.

1) promise	a) adversity	b) potential	c) unification
2) uncommitted	a) free	b) organized	c) fresh
3) dilemma	a) problem	b) tissue	c) garden
4) harvested	a) adjusted	b) replaced	c) taken
5) opposed	a) negative	b) against	c) pro
6) breakthrough	a) discovery	b) reality	c) attempt

CROSSWORD PUZZLE

ACROSS:

1) The tissue will be a perfect ____________________ with the DNA of the patient.
2) Adult stem cells are ____________________ for scientists to find because there are such a small number of them.
4) What is a stem ____________________?
5) They want to be able to grow the cells into whatever type of replacement ____________________ is needed.
9) ____________________ hope that someday stem cells can be taken directly from a sick adult person and grown into the tissue that person needs to help them recover.
11) Surprisingly, stem cells do ____________________ in adults.
12) Stem cell cloning involves an embryo with the same genetic information as the ________________.

DOWN:

1) A breakthrough using stem cells could change ____________________ forever.
3) Scientists believe in the ____________________ they will find a way.
6) Their opposition can be on moral, ____________________ or religious grounds.
7) Scientists grow replacement ____________________ tissue for you.
8) At present, there are ____________________ kinds of stem cell research.
9) Wouldn't it be great if doctors could ____________________ worn out cells in your body?
10) Many stem cells can be found in a human ____________________ in its early stages of development.

ANSWER KEY

THE MAIN IDEA

c) An explanation of stem cells and stem cell research.

UNDERSTANDING WHAT YOU READ

1) The promise of stem cell research is that someday scientists may be able to cure many diseases and afflictions by growing replacement parts for the sick person.
2) The simplest way to put it is a stem cell is a human cell that has yet to make up its mind to which part of the body it belongs. The cells have yet to become specialized.
3) Most of the cells in an adult are already part of an organ or a nerve or some other part of the body.
4) Many people are opposed to stem cell research. Their opposition can be on moral, ethical or religious grounds.

REMEMBERING DETAILS

1) F There are more stem cells in embryos than there are in adults.
2) T
3) F Like with embryonic and adult stem cell research, scientists are still looking for answers how to control the stem cells they get through cloning.
4) F A breakthrough using stem cells could change medicine forever and offer humans cures for many of the physical ills that strike us.

INFERENCES

1) a
2) a

WORD POWER

1) b
2) a
3) a
4) c
5) b
6) a

CROSSWORD PUZZLE

ACROSS: 1) match 2) difficult 4) cell 5) tissue 9) researchers 11) exist 12) patient

DOWN: 1) medicine 3) future 6) ethical 7) brain 8) three 9) replace 10) embryo

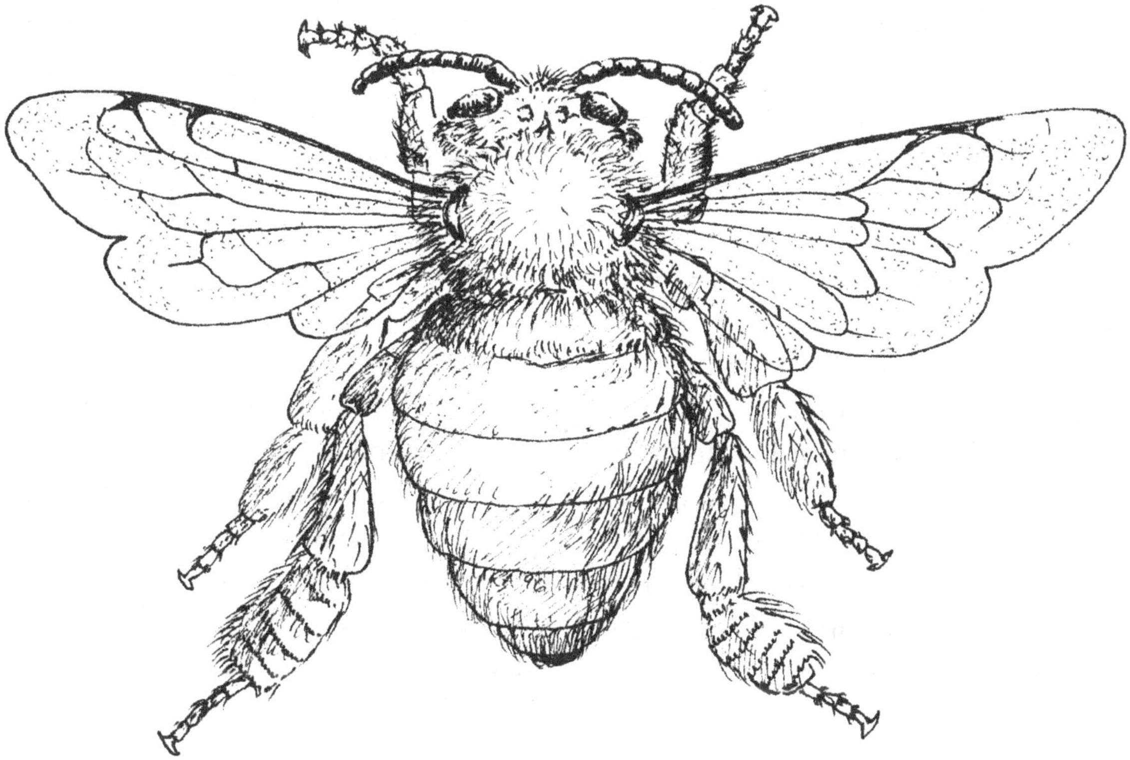

Killer Bees

1 A killer bee looks almost exactly the same as any other honeybee. Actually, entomologists, scientists who study insects, call them African bees. The term killer bees became popular because of a Hollywood movie, *The Swarm*, and because of a few instances where people were actually killed by swarming African bees. Everyone has heard about how ferocious killer bees are. Killer bees arrived in the United States in 1990.

2 African bees are different from most other bees because of the climate in Africa. With no prolonged cold season, African bees need not worry about collecting supplies for winter; therefore, they build smaller nests. Because they build smaller nests, overcrowding becomes a problem. At such times, some of the bees leave to form a new colony. New colonies are always forming and migrating. When the bees move together like this, it is called swarming. African bees can swarm a few times a year, while bees from more northern regions rarely do. Because it never gets cold, African bees can change nests if their food supply becomes unsatisfactory or if they sense any danger to their nest.

African bees are more hostile than 3
other bees because African bees have always built their nests out in the open, so they have been vulnerable to attack. African bees are now instinctively fierce. If one bee senses danger, it can send a signal that will have thousands of bees ready to attack in a matter of seconds. Swarming killer bees have

been credited with killing an animal as large as a horse, and one man who was killed is said to have had suffered over eight thousand stings.

4 African bees were introduced to the Americas in Brazil in 1956. They were imported into Brazil as an experiment to help boost national honey production. The experiments were supposed to be held under the strictest of conditions to prevent any queen bees from escaping. The scientists thought it could never happen, but it did. A queen and her followers escaped. Killer bees were loose in Brazil.

5 And it did not take long for African bees to take root in Brazil. The newcomers felt right at home. New colonies spread quickly as the bees reproduced and swarmed. African male bees mated with local queen bees, and their offspring were aggressive like their fathers. African bees attacked nests, killed other bees and stole honey. Word of attacks on people by hundreds of bees were reported by the media, and news of the killer bees spread, as did the bees. By the early 1980s they had reached Panama and by the mid-1980s they had reached Mexico.

The killer bees arrived in Texas, 6
U.S.A. in 1990. Today African bees have been spotted in all the most southern states. So far it remains to be seen how far north the bees will be able to go and to survive. Some scientists think killer bees will be confined to the lower states where it remains relatively warm all year, while other scientists think the bees will make it all the way to Canada. Time will tell.

One thing must be understood about 7
killer bees: they do not attack without a reason. If you live in an area where killer bees are known to exist, be cautious around all bees. Do not disturb any bees' nest. If you live in an area where killer bees do not currently exist, be happy for now, because they may be arriving soon.

THE MAIN IDEA

Circle the letter of the sentence or phrase which best describes the main idea of the article about killer bees. Be prepared to support your answer.

a) The threat from killer bees.
b) The facts about killer bees.
c) The origin of killer bees.
d) Being safe around killer bees.

UNDERSTANDING WHAT YOU READ

If you can, answer these questions from memory. If you cannot, look back at the article.

1) Why did the term killer bees become popular?

____

2) Why do African bees swarm more often than other types of honeybees?

____

3) In what year were African bees imported into Brazil?

____

4) How far north do scientists think killer bees will be able to go and to survive?

____

REMEMBERING DETAILS

Write TRUE or FALSE under each statement. If the statement is false, write the statement correctly.

1) It is only since being imported to the Americas that African bees have begun building their nests out in the open.

____

2) The African bees had problems adjusting to life in Brazil.

____

3) The media spread the killer bees.

____

4) Killer bees do not attack without a reason.

____

INFERENCES

Based on the article, circle the letter of the best sentence completion.

1) African bees were brought to Brazil to...

a) help honey farmers.
b) replace the existing bee population.
c) introduce beekeeping and making honey as a new industry there.
d) mate with the local bee population to produce more aggressive offspring.

2) We can conclude that the spread of killer bees...

a) is guaranteed to remain only in the southern states.
b) could go as far north as Canada; nobody knows.
c) is not important.
d) has been determined exactly.

INTERPRETATION

1) Are there bees in your country? Are they dangerous or helpful? Explain your answer.

2) Killer bees are non-native creatures that have been introduced to the Americas. Can you think of other examples where non-native creatures were brought into a new environment? What were the results?

3) Killer bees spread because scientist were not careful. What is the responsibility of scientists in such cases? Can you think of other examples where science has had an unintended effect?

WORD POWER

Circle the letter of the word that means the same as the word on the left.

1) ferocious	a) timid	b) dedicated	c) fierce
2) climate	a) elevation	b) weather	c) rainfall
3) vulnerable	a) open	b) expected	c) asking
4) strictest	a) tightest	b) least severe	c) highest
5) spotted	a) targeted	b) sprayed	c) seen
6) cautious	a) careful	b) daring	c) clumsy

CROSSWORD PUZZLE

ACROSS:

1) Killer bees ____________________ in the United States in 1990.

3) Do not ____________________ any bees' nest.

5) African bees can ____________________ a few times a year.

6) Some scientists think killer bees will be ____________________ to the lower states.

8) ____________________ are scientists who study insects.

12) A killer bee looks almost exactly the same as any other ____________________.

DOWN:

1) One signal can have thousands of bees ready to ____________________ in a matter of seconds.

2) African bees are more ____________________ than other bees.

4) One man who was killed is said to have had suffered over eight thousand ____________________.

7) A queen and her ____________________ escaped.

9) The ____________________ felt right at home in Brazil.

10) New colonies are always forming and ____________________.

11) By the early 1980s they had ____________________ Panama.

13) African bees were imported to help ____________________ national honey production.

ANSWER KEY

THE MAIN IDEA

b) The facts about killer bees.

UNDERSTANDING WHAT YOU READ

1) The term killer bees became popular because of a Hollywood movie, *The Swarm*, and because of a few instances where people were actually killed by swarming African bees.
2) African bees swarm more often than other types of bees because the African climate allows them to, and because African bees build small nests overcrowding leads them to swarm more often.
3) African bees were imported into Brazil in 1956.
4) Scientists still debate how far north killer bees will be able to go and to survive. Some scientists think killer bees will be confined to the lower states where it remains relatively warm all year, while other scientists think the bees will make it all the way to Canada.

REMEMBERING DETAILS

1) F African bees have always built their nests out in the open.
2) F It did not take long for African bees to take root in Brazil. The newcomers felt right at home. New colonies spread quickly as the bees reproduced and swarmed.
3) F Word of attacks on people by hundreds of bees were reported by the media, and news of the killer bees spread. The media did not spread the bees.
4) T

INFERENCES

1) a
2) b

WORD POWER

1) c
2) b
3) a
4) a
5) c
6) a

CROSSWORD PUZZLE

ACROSS: 1) arrived 3) disturb 5) swarm 6) confined 8) entomologists 12) honeybee

DOWN: 1) attack 2) hostile 4) stings 7) followers 9) newcomers 10) migrating 11) reached 13) boost

Alzheimer's Disease

1 Alzheimer's disease (AD) is the most common cause of intellectual deterioration in middle-aged and elderly North Americans, affecting more than two and one-quarter million men and women over the age of 65. The usual symptoms of AD are memory loss and a progressive loss of mental faculties. Presently, there is no cure for Alzheimer's disease.

2 The human brain is a complex organ that acts like a main control room for the rest of the body. It gathers information, controls body movements, and activates emotions and thoughts. The brain does this by using a chemical called acetylcholine to send signals to and from brain cells. Alzheimer's disease causes a breakdown in the neurons of brain cells by creating a shortage of acetylcholine.

Once a person is affected by Alzheimer's, it becomes more difficult for the brain to perform its main functions. These changes in the brain are so intense and crippling that a person diagnosed with Alzheimer's disease usually dies within 10 years. 3

For a long time the symptoms of Alzheimer's disease (memory loss, slurred speech, and confusion) were thought to be the natural side effects of getting old. AD went unrecognized because victims often seemed physically healthy; it was just their memory and general mental performance that were affected. 4

5 In 1906, Alois Alzheimer, a German neuropathologist, discovered two abnormalities in the brain of a fifty-six year old woman who died of severe mental problems. For a long time Alzheimer's discovery was overlooked because doctors of that era (and for many years afterwards) thought these irregularities were only the normal consequences of human aging.

6 In the mid-1980s Alzheimer's theories were re-examined and AD came to be recognized as the most common cause of mental decay in the elderly. In fact, the many problems related to AD are considered so great that Alzheimer's disease has been called "the disease of the 20th century" by doctor and author Lewis Thomas.

7 Although the cause of Alzheimer's disease is still unknown, two risk factors have been identified: advanced age and genetic history. The risk of developing AD is less than one percent before the age of 50, but the risk increases as a person gets older. If a person reaches age 90, they have a 30 percent chance of developing Alzheimer's disease.

In families that already have a member with Alzheimer's disease, relatives have a 50 percent chance of developing AD because the tendency is passed on as a genetic trait. Recent research has discovered a consistent abnormal gene site on chromosome 21 of people affected with Alzheimer's disease. Scientists are hopeful that this finding is a step in the direction of a cure. 8

Because the exact cause of Alzheimer's disease has not yet been determined, an effective treatment or a cure is still not available. Present treatments include drugs to relieve patients' symptoms. Support groups and counseling are also available for families living with AD. Scientists and medical researchers are optimistic that the same medical advances that are now extending life for everyone will soon help eliminate Alzheimer's disease as well. 9

THE MAIN IDEA

Circle the letter of the sentence or phrase which best describes the main idea of the article about Alzheimer's disease. Be prepared to support your answer.

a) The main symptoms of Alzheimer's disease.
b) The "disease of the 20th century."
c) The causes and effects of Alzheimer's disease.
d) Factors that increase the risk of Alzheimer's disease.

UNDERSTANDING WHAT YOU READ

If you can, answer these questions from memory. If you cannot, look back at the article.

1) In North America, how common is Alzheimer's disease?

---

2) How does the human brain function?

---

3) How does Alzheimer's disease affect the brain?

---

4) Which two groups are at most risk to develop Alzheimer's disease?

REMEMBERING DETAILS

Write TRUE or FALSE under each statement. If the statement is false, write the statement correctly.

1) The original discovery of AD was not properly understood because doctors just thought it was normal for old people to have mental problems.

---

2) A good diet and lots of exercise will reduce the risk of Alzheimer's disease.

---

3) Alzheimer's disease was called "the disease of the 20th century" because it was first discovered early in this century.

---

4) Doctors hope to find a clue that may help explain this disease.

INFERENCES

Based on the article, circle the letter of the best sentence completion.

1) It seems that knowing about Alzheimer's disease may help everyone...

a) look after young children better.
b) provide better care for people under 50.
c) meet the medical requirements of families.
d) understand the needs of old people.

2) Medical treatment for Alzheimer's disease...

a) is only useful if the disease is detected early.
b) is only effective for some people.
c) is still experimental.
d) has finally eliminated this terrible illness.

INTERPRETATION

1) When you think of the danger of age-related diseases like AD, do you think that it is wise for doctors to look for ways to help people live longer?

2) How are elderly people treated in your culture? How do different generations relate to each other?

3) Think of one very old person you have known. How has that person affected your life, and what lessons has that person taught you?

WORD POWER

Circle the letter of the word that means the same as the word on the left.

1) deterioration	a) administration	b) decay	c) heightening
2) symptoms	a) signs	b) devices	c) pains
3) shortage	a) too much	b) just enough	c) too little
4) slurred	a) unclear	b) slow	c) not grammatical
5) consistent	a) important	b) thick	c) constant
6) optimistic	a) worried	b) hopeful	c) excited

CROSSWORD PUZZLE

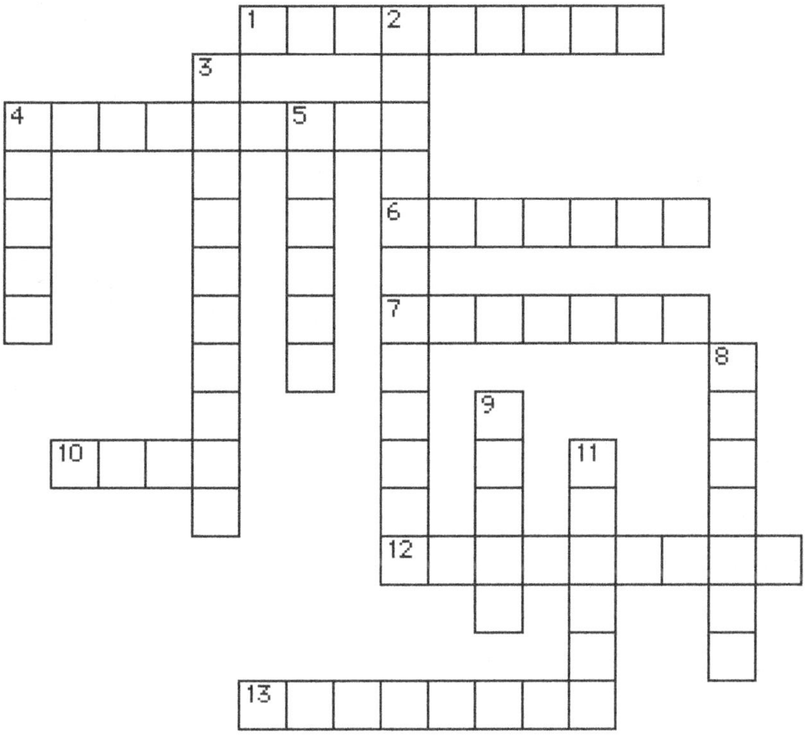

ACROSS:

1) The symptoms of AD are memory loss and a progressive loss of mental ____________________.

4) Alois ____________________ discovered two abnormalities in the brain of a fifty-six year old.

6) AD has been called "the disease of the ____________________."

7) Alzheimer's disease is passed on as a ____________________ trait.

10) Scientists are hopeful that they are moving in the direction of a ____________________.

12) A person ____________________ with Alzheimer's disease usually dies within 10 years.

13) Recent research has discovered a consistent ____________________ gene site on a chromosome.

DOWN:

2) AD went ____________________ because victims often seemed physically healthy.

3) The exact cause of Alzheimer's disease has not yet been ____________________.

4) Doctors thought these irregularities were the normal consequences of ____________________.

5) A person's ____________________ and their general mental performance are affected.

8) The risk of developing AD is less than one ____________________ before the age of 50.

9) The human brain is a complex ____________________.

11) AD is recognized as the most common cause of ____________________ decay in the elderly.

ANSWER KEY

THE MAIN IDEA

c) The causes and effects of Alzheimer's Disease.

UNDERSTANDING WHAT YOU READ

1) Alzheimer's disease is the most common form of brain disorder in North America, affecting more than two and one-quarter million men and women over the age of 65.
2) The brain functions like a main control room for the rest of the body. It gathers information, controls movements, and activates thoughts and emotions. It communicates with the body by using the chemical acetylcholine to send signals to and from brain cells.
3) Alzheimer's disease affects the brain by causing a breakdown in the neurons of brain cells, so that not enough acetylcholine is produced.
4) The two groups which are at most risk to develop Alzheimer's disease are elderly people and people who have close relatives with the disease.

REMEMBERING DETAILS

1) T
2) F Age and family background seem to be what determine the risk.
3) F Yes, it was discovered in 1906, but the real reason for the title, "disease of the 20th century" is that AD became very important in the last century, when people were living longer and so many people were affected by it.
4) F In fact, doctors have already found a clue: an abnormal gene in people who have Alzheimer's disease.

INFERENCES

1) d
2) c

WORD POWER

1) b
2) a
3) c
4) a
5) c
6) b

CROSSWORD PUZZLE

ACROSS: 1) faculties 4) Alzheimer 6) century 7) genetic 10) cure 12) diagnosed 13) abnormal

DOWN: 2) unrecognized 3) determined 4) aging 5) memory 8) percent 9) organ 11) mental

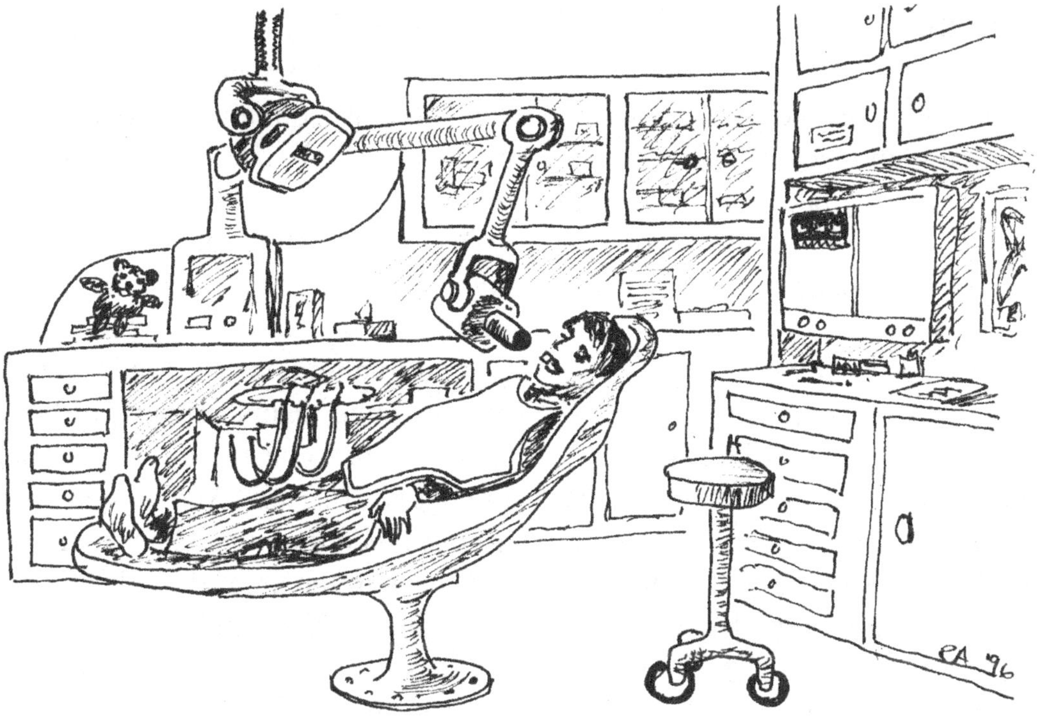

Dentistry

1 A lot happens to your teeth. They are ground, braced, brushed, flossed, polished, drilled, filled and, if you do not take care of them, pulled. For your teeth, proper oral care is as essential as anything else you do for your body and your general health. Fortunately, if there is any damage to your teeth modern dentistry techniques make it easier, faster and less painful to get the problem fixed.

2 By definition, dentistry includes the study, diagnosis, treatment and the prevention of diseases of the mouth, teeth, gums and jawbones. Today's dentists use sophisticated equipment and pain-preventing medicines, but these have not always been around.

3 The earliest historical records of dentistry are from Egypt from around 3600 BC. Aesculapius, a Greek physician who lived around 1250 BC, is credited with the idea of pulling diseased teeth. Hippocrates, around 400 BC, wrote of applying ointments and using branding irons to treat diseases of the mouth. Through the Middle Ages wealthy people were treated by physicians and surgeons, while poor people could look to the local barber to pull painful teeth.

Modern dentistry emerged with the release of Pierre Fauchard's book *The Surgeon Dentist, A Treatise on Teeth.* Fauchard (1678-1761) is considered the father of modern dentistry. Later, the first school devoted to teaching dentistry opened in Baltimore in the USA in 1840. 4

5 Greene Vardiman Black (1831-1915) invented a foot engine so dentists would have two free hands while using the dental drill. He also came up with the idea that gum disease was an infection caused by bacteria. This was later confirmed in 1960. An important contribution was made in 1944 when Dr. Horace Wells, a Connecticut dentist, first used nitrous oxide (laughing gas) as an oral anesthetic; today this method of administering anesthetic is employed for many dental and medical surgeries. The introduction of fluoride in toothpaste in the 1960s was a huge breakthrough; teenagers and adults today have a 50% better chance of not developing tooth decay.

6 The 1990s saw a change in thinking when it came to fillings for cavities in teeth. Since the early 1800s silver amalgam--a combination of silver, tin and mercury--has been the standard filling compound. But modern fears about mercury poisoning have led many people to have their amalgam fillings replaced with plastic polymer fillings. Besides not having any mercury, plastic polymers have an advantage over metal fillings because the color can be matched to the surrounding tooth and the material can be chemically bonded to the tooth.

Dentistry methods are sure to advance even more in the future. 7 Things such as laser technology, which is used to eliminate diseased tissue or weld together damaged parts of teeth, will soon become a common practice in dentists' offices. And computer-assisted technology is ushering in new advances in the area of tooth restoration. The primary goal of dentists is to develop early detection methods to prevent dental conditions from turning into dental problems.

Caring for your teeth properly will 8 lessen the number of visits you make to the dentist. Brushing properly, flossing every day, getting a professional cleaning every six months and maintaining a diet low in sugar can aid in keeping your teeth healthy. Your teeth will stay in your mouth if you take care of them!

THE MAIN IDEA

Circle the letter of the sentence or phrase which best describes the main idea of the article about dentistry. Be prepared to support your answer.

a) Proper dental hygiene.
b) Major events in the history of dentistry.
c) Great people who have been dentists.
d) Dentistry methods.

UNDERSTANDING WHAT YOU READ

If you can, answer these questions from memory. If you cannot, look back at the article.

1) Who was Aesculapius? What did he do?

---

2) Who was Pierre Fauchard?

---

3) What did Dr. Horace Wells do?

---

4) What are some good ways to aid in keeping your teeth healthy?

REMEMBERING DETAILS

Write TRUE or FALSE under each statement. If the statement is false, write the statement correctly.

1) Modern dentistry techniques are just as painful as techniques were in the past.

---

2) Pierre Fauchard is considered the father of modern dentistry.

---

3) Dr. Horace Wells came up with the idea that gum disease was an infection caused by bacteria.

---

4) The primary goal of dentists is to treat dental problems.

INFERENCES

Based on the article, circle the letter of the best sentence completion.

1) Dr. Horace Wells' discovery...

a) led to the introduction of fluoride.
b) was less important than Greene Vardiman Black's foot engine.
c) left many patients laughing.
d) had a much wider impact than just on dentistry.

2) Teenagers and adults have had a 50% better chance of not developing tooth decay since...

a) they started brushing their teeth more.
b) the 1960s.
c) amalgam fillings have been replaced by plastic polymer.
d) Pierre Fauchard's book was released.

INTERPRETATION

1) Would you like to be a dentist? Why or why not? What would it be like?

2) How do you feel about going to visit the dentist? Have your experiences been positive or negative? Give examples.

3) With a partner, develop a short role play in which one of you is the dentist and the other is the patient. Try to incorporate something unusual into your skit.

WORD POWER

Circle the letter of the word that means the same as the word on the left.

1) credited	a) invented	b) acknowledged	c) insured
2) chance	a) possibility	b) leeway	c) border
3) cavities	a) holes	b) mounds	c) canker
4) bonded	a) blended	b) booked	c) cemented
5) ushering	a) leading	b) sending	c) plugging
6) detection	a) discovery	b) eyeing	c) extraction

CROSSWORD PUZZLE

ACROSS:

1) Hippocrates wrote of using ____________________ irons to treat diseases of the mouth.

6) The introduction of fluoride in ____________________ in the 1960s was a huge breakthrough.

8) Today's dentists use ____________________ equipment and pain-preventing medicines.

11) Using ____________________ improves your chances of not developing tooth decay.

12) Many people are having their old fillings replaced with plastic ____________________ fillings.

13) G.V. Black realized that ____________________ disease was an infection caused by bacteria.

DOWN:

2) Computers are ushering in new advances in the area of tooth ____________________.

3) The earliest historical records of ____________________ are from Egypt from around 3600 BC.

4) A lot happens to your ____________________.

5) Your teeth will stay in your ____________________ if you take care of them.

7) Teenagers and adults today have a better chance of not developing tooth ____________________.

9) The first school ____________________ to teaching dentistry opened in Baltimore in 1840.

10) Poor people could look to the local ____________________ to pull painful teeth.

11) Since the early 1800s silver amalgam has been the standard ____________________ compound.

ANSWER KEY

THE MAIN IDEA

b) Major events in the history of dentistry.

UNDERSTANDING WHAT YOU READ

1) Aesculapius was a Greek physician who lived around 1250 BC. He is credited with the idea of pulling diseased teeth.
2) Pierre Fauchard (1678-1761) is considered the father of modern dentistry. Modern dentistry emerged with the release of his book *The Surgeon Dentist, A Treatise on Teeth.*
3) Dr. Horace Wells, a Connecticut dentist, first used nitrous oxide (laughing gas) as an oral anesthetic in 1944. It was an important contribution to dentistry, and today this method of administering anesthetic is employed for many dental and medical surgeries.
4) Brushing properly, flossing every day, getting a professional cleaning every six months and maintaining a diet low in sugar can aid in keeping your teeth healthy.

REMEMBERING DETAILS

1) F Modern dentistry techniques are less painful than techniques were in the past.
2) T
3) F Greene Vardiman Black came up with the idea that gum disease was an infection caused by bacteria.
4) F The primary goal of dentists is to develop early detection methods to prevent dental conditions from turning into dental problems.

INFERENCES

1) d
2) b

WORD POWER

1) b
2) a
3) a
4) c
5) a
6) a

CROSSWORD PUZZLE

ACROSS: 1) branding 6) toothpaste 8) sophisticated 11) fluoride 12) polymer 13) gum

DOWN: 2) restoration 3) dentistry 4) teeth 5) mouth 7) decay 9) devoted 10) barber 11) filling

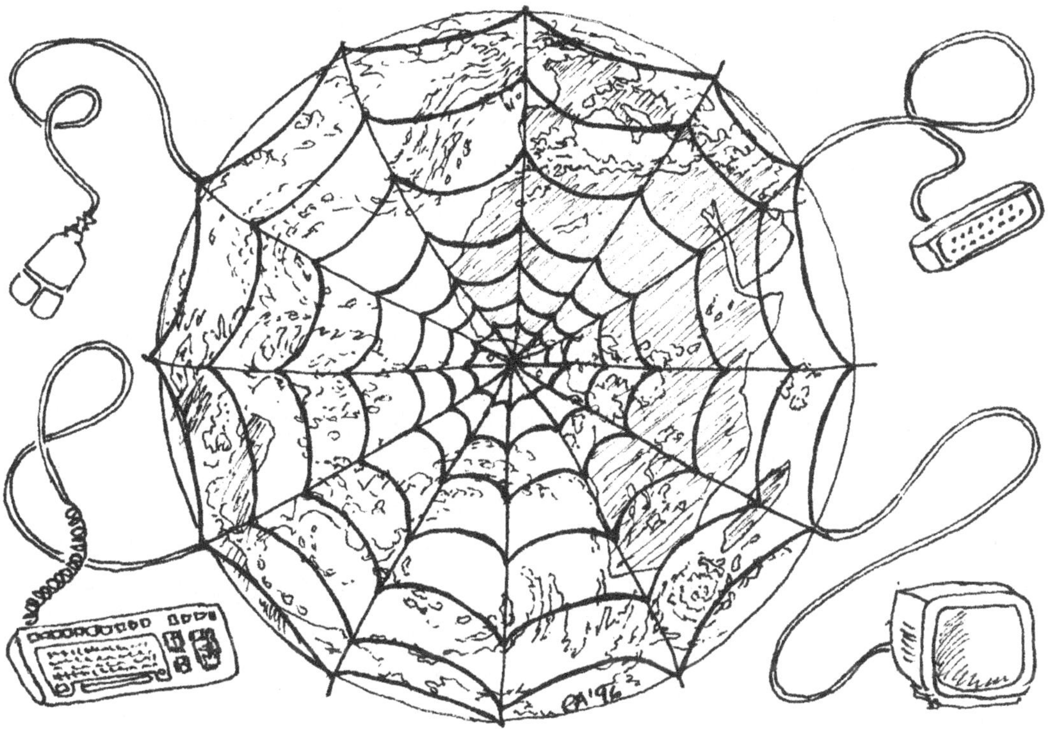

Internet

1 The Internet is a worldwide network of computers. In fact, the Internet is made up of thousands of networks of computers that have the ability to exchange information. Other names you may have heard for the Internet are Cyberspace and the Information Superhighway. Here are a few things you might like to know before you go "surfing" the Internet.

2 The Internet started as a test program by the United States Defense Department. This was in 1969. The military wanted to design a computer network where there was no one main computer, and where every computer in the system could "talk" with every other computer. That way, in case part of the network was destroyed by an enemy attack, communications would still be possible because the surviving network would still function.

By the mid-1980s, almost 1,000 similar networks had been established. 3
Networks were set up by governments, libraries, businesses and individuals. As more people realized the usefulness of being on the Internet, more and more users came online. By the early 2000s the number of networks is closer to 100,000 and over one billion people are now connected to the Internet. These numbers continue to grow at a staggering rate.

You can find almost anything 4
imaginable on the Internet. You can get information from some of the biggest libraries in the world. You can check out how your favorite sports

team is doing. You can buy all sorts of things, if you go online shopping. You can “download” (electronically copy) data and software from the Internet. You can listen to radio stations that broadcast over the Internet. You can send electronic mail, also called e-mail. You can spend hours visiting countless interesting web pages on the World Wide Web.

5 To get connected to the Internet you need the following equipment: a computer, a modem, and the special software it takes to use the Internet. You must also pay a company, called a service provider, for access to the Internet, usually through a phone line or through a cable television hook-up.

6 E-mail is the most popular part of the Internet, probably because users can avoid most long distance telephone charges by using e-mail rather than talking. With e-mail it takes seconds to send a message anywhere in the world. If you are not home to receive an e-mail message, your service provider will keep it for you until you retrieve it. There is easy to use software to send and receive e-mail messages, and you have a unique e-mail address to identify your mailbox.

7 The World Wide Web (WWW) is one particular system on the Internet. Vast amounts of information are available on the WWW. The documents inside the WWW are called Web pages. Web pages are stored at places called Web sites, which can be a company, or a university or any agency that stores Web pages. When you explore the WWW you use a special program called a Web browser. All Web pages have a unique address known as the Uniform Resource Locator, or URL.

8 By allowing the nearly instantaneous transfer of information to almost anywhere on the planet, the Internet has helped to bring neighbors in the global village even closer together.

THE MAIN IDEA

Circle the letter of the sentence or phrase which best describes the main idea of the article about the Internet. Be prepared to support your answer.

a) The value of the Internet.
b) The effect of the Internet on the global village.
c) E-mail and the World Wide Web.
d) History and features of the Internet.

UNDERSTANDING WHAT YOU READ

If you can, answer these questions from memory. If you cannot, look back at the article.

1) How did the Internet start?

---

2) By the early 2000s, how people were connected to the Internet?

---

3) What equipment do you need to get connected to the Internet?

---

4) What happens if you are not home to receive an e-mail message?

REMEMBERING DETAILS

Write TRUE or FALSE under each statement. If the statement is false, write the statement correctly.

1) The military wanted to design a computer network where one main computer controlled all information exchanges.

---

2) In recent years, the number of new users going online has slowed considerably.

---

3) You must pay a service provider to get access to the Internet.

---

4) All e-mail mailboxes have a Uniform Resource Locator.

INFERENCES

Based on the article, circle the letter of the best sentence completion.

1) The number of Internet users is growing quickly because...

a) service providers have lowered their prices.
b) people are realizing that the Internet has a lot to offer.
c) using the Internet is a trend that is going to end quickly.
d) people want to find out what the U.S. Defense Department is doing.

2) To use the Internet and its features, it seems you need...

a) fast hardware.
b) special software.
c) to understand how to "download" data.
d) friends in other parts of the world.

INTERPRETATION

1) Do you use the Internet? If so, what do you do? If not, would you like to try? Why or why not?

2) Think of the different things you do each day that involve computers in some way. How have computers made your life easier?

3) What do you think will be the future impact of the Internet on our daily lives? Give examples. Will the effect be good or bad?

WORD POWER

Circle the letter of the word that means the same as the word on the left.

1) main	a) alien	b) central	c) compatible
2) function	a) work	b) buzz	c) switch
3) staggering	a) astounding	b) lamentable	c) depressing
4) countless	a) relentless	b) surprisingly	c) numerous
5) retrieve	a) transmit	b) collect	c) detain
6) instantaneous	a) immediate	b) spontaneous	c) regrettable

CROSSWORD PUZZLE

ACROSS:
2) You can buy all sorts of things, if you go ____________________ shopping.
3) The Internet is a network of computers that have the ability to ____________________ information.
4) The ____________________ network would still function.
6) Other names for the Internet are ____________________ and the Information Superhighway
9) Vast amounts of ____________________ are available on the World Wide Web.
12) The Internet started as a test ____________________ by the United States Defense Department.
13) By the mid-1980s, almost 1,000 ____________________ networks had been established.

DOWN:
1) The ____________________ inside the World Wide Web are called Web pages.
3) You can send ____________________ mail, also called e-mail.
5) The Internet has helped to bring ____________________ in the global village even closer together.
7) You must pay a company, called a service ____________________, for access to the Internet.
8) You can get information from some of the ____________________ libraries in the world.
10) With e-mail it takes seconds to send a ____________________ anywhere in the world.
11) E-mail is the most ____________________ part of the Internet.

ANSWER KEY

THE MAIN IDEA

d) History and features of the Internet.

UNDERSTANDING WHAT YOU READ

1) The Internet started as a test program by the United States Defense Department.
2) By the early 2000s over one billion people were connected to the Internet.
3) To get connected to the Internet you need the following equipment: a computer, a modem, and the special software it takes to use the Internet. You must also pay a company, called a service provider, for access to the Internet.
4) If you are not home to receive an e-mail message, your service provider will keep it for you until you retrieve it.

REMEMBERING DETAILS

1) F The military wanted to design a computer network where there was not one main computer that controlled all information exchanges.
2) F In recent years, the number of new users going online has continued to rise at a staggering rate.
3) T
4) F An e-mail mailbox has an address, but it is a Web page that has a Uniform Resource Locator.

INFERENCES

1) b
2) b

WORD POWER

1) b
2) a
3) a
4) c
5) b
6) a

CROSSWORD PUZZLE

ACROSS: 2) online 3) exchange 4) surviving 6) Cyberspace 9) information 12) program 13) similar

DOWN: 1) documents 3) electronic 5) neighbors 7) provider 8) biggest 10) message 11) popular

ØUB3

1 In 2005 something happened for the first time since 1930. Pluto, at that time the 9th planet in our solar system, was discovered in 1930. In early 2005 three astronomers working near San Diego, California noticed something fascinating on photographs of the night sky. The photographs had been taken by the planetary scientists in 2003. What had Michael Brown, Chad Trujillo and David Rabinowitz found?

2 When they made their find public the planetoid was officially dubbed 2003 UB313. However, rumor had it that the three happy scientists called their new celestial body Xena, after the heroine on a popular television show. Many people, including the discoverers, NASA and many people in the media, claimed this to be the discovery of a 10th planet. But would Xena become the tenth planet?

2003 UB313 is the farthest known object in the solar system. It is 15 billion kilometers from the Sun. It is almost 100 times as far from the Sun as is the Earth. One orbit of the Sun takes 2003 UB313 over five hundred and fifty years. The surface of Xena is thought to be covered in methane gas and methane ice. There is a moon which orbits 2003 UB313. 3

Measurements of the size of 2003 UB313 taken by the Hubble Space Telescope show that it is bigger than Pluto, almost 5 percent bigger. If 2003 UB313 is bigger than Pluto and Pluto was a planet then 2003 UB313 should be a planet too, right? Well, that depended on a group known as the 4

International Astronomical Union (IAU).

5 The IAU is an organization of professional astronomers from around the world. The IAU's job is to monitor and oversee all things pertaining to astronomy. The IAU judges on things like whether 2003 UB313 gets to be called a planet. 2003 UB313 is bigger than Pluto. 2003 UB313 is a planet, right? There was one big problem. The IAU had to make a decision on the definition of the word planet.

6 The dictionary gives a definition of planet, but in fact not so long ago there was no one accepted scientific definition of the word. IAU astronomers declared that they would publish the definition of what is a planet in August 2006. Their decision would decide whether 2003 UB313 would become the tenth planet. It has been reported that the IAU voted to decide on using one of the following three definitions: 1) An object with a diameter greater than 2000 km. 2) An object whose shape is stable due to its own gravity. 3) An object that dominates its district.

7 Using the first definition 2003 UB313 would become recognized as the tenth planet in our solar system. Using the second definition many objects not now considered planets would be considered planets. The third definition would reduce the number of planets in our solar system to eight. 2003 UB313 would not become the tenth planet and Pluto would lose its status as a major planet.

8 The decision the IAU made was to reduce the number of planets to eight. 2003 UB313 did not become the tenth planet. Pluto lost its status as a planet. One thing is certain their decision changed the universe!

THE MAIN IDEA

Circle the letter of the sentence or phrase which best describes the main idea of the article about 2003 UB313. Be prepared to support your answer.

a) The controversy surrounding a definition of the word planet.
b) The story of 2003 UB313.
c) The workings of the International Astronomical Union.
d) The number of planets in our solar system.

UNDERSTANDING WHAT YOU READ

If you can, answer these questions from memory. If you cannot, look back at the article.

1) When was 2003 UB313 photographed? When was it discovered?

---

2) How far is 2003 UB313 from the Sun? How much father than Earth is 2003 UB313 from the Sun?

---

3) What is the IAU? What's its job?

---

4) Explain how the three proposed definitions would change the solar system.

REMEMBERING DETAILS

Write TRUE or FALSE under each statement. If the statement is false, write the statement correctly.

1) 2003 UB313 is the farthest known object in the solar system.

---

2) 2003 UB313 is five percent smaller than Pluto.

---

3) Before August 2006 there was no definition of planet in the dictionary, but there was a scientific definition.

---

4) The decision as to whether Xena became the tenth planet was in the hands of Michael Brown, Chad Trujillo and David Rabinowitz.

INFERENCES

Based on the article, circle the letter of the best sentence completion.

1) The three scientists were probably...

a) hoping the first definition won the vote.
b) hoping the second definition won the vote.
c) hoping the third definition won the vote.
d) had no real interest in the vote because they believe what they found is a planet.

2) It seems strange that...

a) new planets are being discovered.
b) the scientists named their planetoid after a television character.
c) 2003 UB313 is bigger than Pluto.
d) before 2006 no scientific definition of planet had been accepted.

INTERPRETATION

1) Hold a vote for the definition of the word planet in your class. Count the votes.

2) For which of the three definitions did you vote? Explain why you chose one over the others.

3) How will the IAU's decision change the universe? Or will it change the universe?

WORD POWER

Circle the letter of the word that means the same as the word on the left.

1) dubbed	a) shouted	b) named	c) marveled
2) media	a) people on street	b) religious people	c) news people
3) pertaining to	a) meddling with	b) relating to	c) clearly as
4) accepted	a) argued	b) withheld	c) held
5) stable	a) steady	b) wavering	c) leaning
6) status	a) market	b) pseudonym	c) standing

CROSSWORD PUZZLE

ACROSS:

1) There is a ____________________ which orbits 2003 UB313.

2) Using the second definition many objects not now considered ____________________ would be considered planets.

3) In 2005 something happened for the first ____________________ since 1930.

4) 2003 UB313 is the ____________________ known object in the solar system.

8) The dictionary gives a ____________________ of planet.

10) One thing is certain their decision changed the ____________________!

11) When they made their find ____________________ the planetoid was dubbed 2003 UB313.

DOWN:

1) The IAU's job is to ____________________ and oversee all things pertaining to astronomy.

5) One orbit of the Sun takes 2003 UB313 over five ____________________ and fifty years.

6) What had Michael Brown, Chad Trujillo and David Rabinowitz ____________________?

7) The third definition would reduce the number of planets in our ____________________ system.

8) The IAU had to make a ____________________ on the definition of the word planet.

9) Their decision decided whether 2003 UB313 became the ____________________ planet.

12) UB313 is ____________________than Pluto.

ANSWER KEY

THE MAIN IDEA

b) The story of 2003 UB313.

UNDERSTANDING WHAT YOU READ

1) The photographs had been taken by the planetary scientists in 2003. In early 2005 they noticed 2003 UB313 on the photographs.
2) It is 15 billion kilometers from the Sun. It is almost 100 times as far from the Sun as is the Earth.
3) The IAU is an organization of professional astronomers from around the world. The IAU's job is to monitor and oversee all things pertaining to astronomy.
4) Using the first definition 2003 UB313 would become recognized as the tenth planet in our solar system. Using the second definition many objects not now considered planets would be considered planets. The third definition would reduce the number of planets in our solar system to eight. 2003 UB313 would not become the tenth planet and Pluto would lose its status as a major planet.

REMEMBERING DETAILS

1) T
2) F 2003 UB313 is five percent bigger than Pluto.
3) F Before August 2006 there was a definition of planet in the dictionary, but there was no scientific definition.
4) F The decision as to whether Xena became the tenth planet was in the hands of the IAU.

INFERENCES

1) a
2) d

WORD POWER

1) b
2) c
3) b
4) c
5) a
6) c

CROSSWORD PUZZLE

ACROSS: 1) moon 2) planets 3) time 4) farthest 8) definition 10) universe 11) public

DOWN: 1) monitor 5) farthest 6) found 7) solar 8) decision 9) tenth 12) bigger

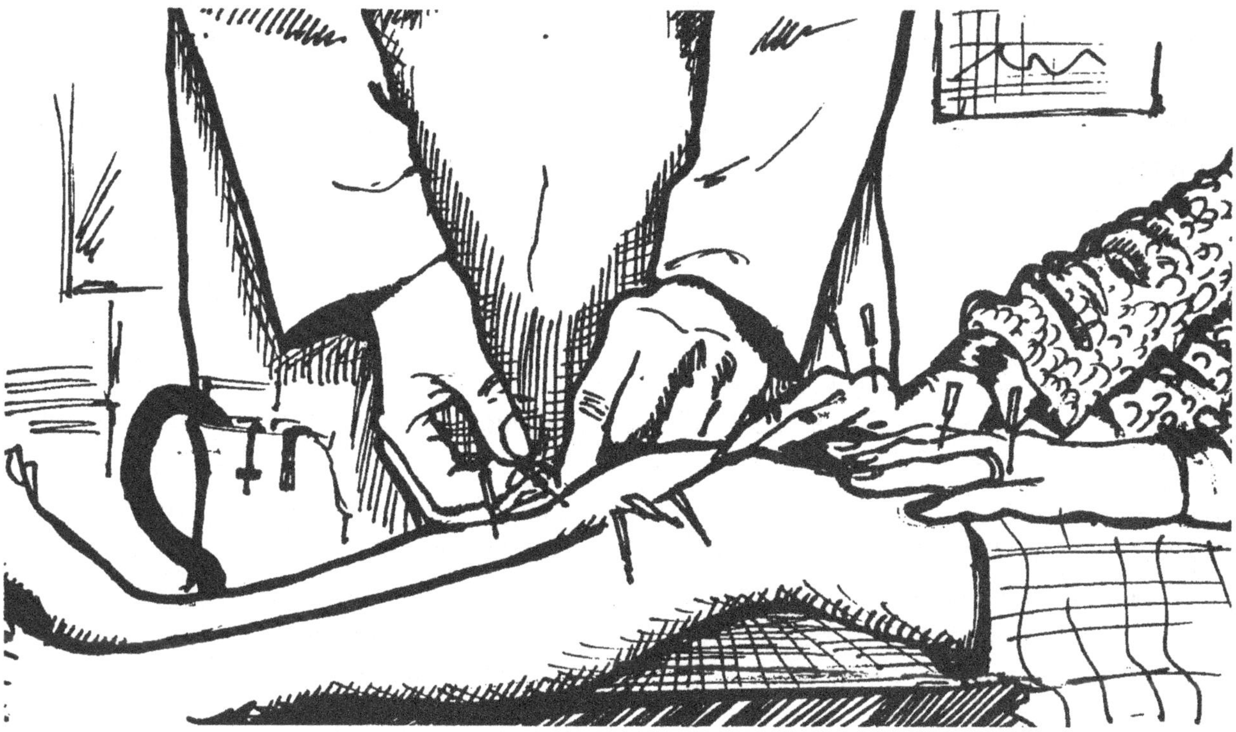

Acupuncture

1 More than 2,000 years ago Chinese physicians discovered they could use needles to treat patients. Thin solid needles are inserted into the body at carefully chosen sites. This practice has long been an important part of ancient Chinese medicine; today acupuncture has earned the respect of doctors in both the East and the West.

2 Ancient Chinese doctors had a well-developed theory of human anatomy. The earliest Chinese anatomical drawing is in a book dating back to the 11th century. This diagram indicates various places where needles can be inserted: acupuncture points. Acupuncturists know which sites on the body are appropriate for the treatment of particular health problems.

According to traditional Chinese acupuncture, our health depends on a good balance of energy circulating in the body. In this view, the human body has a system of pathways through which vital energy flows. If there are blockages along a pathway, or if there is an excess of energy in any one place, the body will experience discomfort or illness. Acupuncture needles are inserted to make the energy flow smoothly again or to reduce pressure. 3

Originally, acupuncture was used only to treat various health problems by restoring a natural flow of energy in the body. In 1966, however, surgeons in China found that they could also employ the technique to prevent pain during operations. Such an anaesthetic use of 4 acupuncture was actually the reverse of the usual approach. Instead of restoring natural body functions, its goal was to upset the body's normal sensation of pain. Nonetheless, the procedure was very effective and at present about a third of all surgery in China is carried out this way.

5 Western interest in acupuncture rose quickly in the 1970s, after its anaesthetic potential became known. From that time on, much research has been carried out in the West as well as the East. In North America, this form of treatment is now used mainly for medical conditions where pain relief is valuable: for example, arthritis, backache or chronic headache. But the modern scientific understanding of acupuncture does not support the old theory of energy levels and pathways. Rather, researchers are studying the possibility that acupuncture may either affect the nervous system, or else perhaps stimulate the release of pain-killing chemicals inside the body.

6 Acupuncture clinics, where therapy is administered by qualified medical practitioners, are now located in most cities of the world. Such treatment is recognized as an effective short-term treatment for various forms of chronic pain. In China, acupuncture anaesthesia is also used during many different kinds of medical operations, including major surgery on the brain, the kidneys, the heart, and the lungs.

7 No doubt the most remarkable quality of acupuncture, as used today, is its capacity to prevent pain without using drugs. Avoiding drugs means avoiding unpleasant side effects. Moreover, there is an opportunity for reduced danger to patients. Ordinary anaesthesia can actually cause death in about one in 3,000 cases; acupuncture anaesthesia does not have this risk. It is interesting to note that, throughout its long history, the key principle in acupuncture has been the treatment of the human body through its own resources, without introducing unnatural chemicals. In fact, this approach is sometimes even used in the treatment of drug addiction!

THE MAIN IDEA

Circle the letter of the sentence or phrase which best describes the main idea of the article about acupuncture. Be prepared to support your answer.

a) Even after more than 2,000 years, the key principle of acupuncture is still the same.
b) Although it is very important in China, acupuncture is not used much in North America.
c) Acupuncture is mostly useful for pain relief.
d) Modern scientific studies have shown that the traditional theory of acupuncture is absolutely true.

UNDERSTANDING WHAT YOU READ

If you can, answer these questions from memory. If you cannot, look back at the article.

1) Where and when did acupuncture originate?

---

2) What do we call the locations where acupuncturists insert needles?

---

3) According to traditional theory, why does the body experience illness?

---

4) What diseases are treated by acupuncture in North America?

REMEMBERING DETAILS

Write TRUE or FALSE under each statement. If the statement is false, write the statement correctly.

1) Thin solid needles are inserted into the body at carefully chosen sites.

---

2) Acupuncture was used only to treat various health problems by causing blockages along a pathway.

---

3) In China, acupuncture anaesthesia is no longer used during medical operations.

---

4) The key principle in acupuncture has been the treatment of the human body by using unnatural chemicals.

INFERENCES

Based on the article, circle the letter of the best sentence completion.

1) Western scientists seem to be coming up with a new explanation of how acupuncture works. This may prove that ancient Chinese doctors...

a) should not have been using acupuncture.
b) were ahead of their time.
c) may have been using acupuncture successfully without understanding how it really works.
d) understood everything about acupuncture.

2) An acupuncture treatment must be given by a trained acupuncture practitioner because...

a) you may be allergic to the drugs a doctor might prescribe.
b) regular doctors do not believe in acupuncture.
c) the needles must be inserted in the right spot to treat each illness.
d) you obviously have a blockage in a pathway.

INTERPRETATION

1) What do you think is the future of acupuncture in North America? Will its use become more widespread or less so? Will acceptance of acupuncture as an alternative to traditional medicines grow or lessen?

2) Have you ever had acupuncture? What were the results? Would you consider acupuncture as an alternative treatment for yourself?

3) Do you understand and accept the theories of acupuncture?

WORD POWER

Circle the letter of the word that means the same as the word on the left.

1) respect	a) attitude	b) denial	c) approval
2) appropriate	a) proper	b) nearby	c) wistful
3) employ	a) cut	b) disregard	c) use
4) chronic	a) recurring	b) temporary	c) mild
5) stimulate	a) return	b) incite	c) detach
6) avoiding	a) keeping away	b) using often	c) taking

CROSSWORD PUZZLE

ACROSS:

3) This approach is sometimes even used in the treatment of drug ____________________!

5) Modern scientific understanding of acupuncture does not support the old theory of ____________________ levels and pathways.

8) Chinese physicians discovered they could use needles to treat ____________________.

11) Our health depends on a good balance of energy ____________________ in the body.

13) Its goal was to upset the body's normal ____________________ of pain.

14) Western ____________________ in acupuncture rose quickly in the 1970s.

DOWN:

1) Acupuncture was used only to treat ____________________ health problems by restoring a natural flow of energy in the body.

2) There is an opportunity for ____________________ danger to patients.

4) Acupuncture anaesthesia is also used during different kinds of medical ____________________.

6) The key principle has been the treatment of the body through its own ____________________.

7) Thin solid needles are ____________________ into the body at carefully chosen sites.

9) If there is an excess of energy in any one place, the body will experience ____________________.

10) The earliest anatomical drawing is in a book ____________________ back to the 11th century.

12) Acupuncture ____________________ are now located in most cities of the world.

ANSWER KEY

THE MAIN IDEA

a) Even after more than 2,000 years, the key principle of acupuncture is still the same.

UNDERSTANDING WHAT YOU READ

1) In China, more than 2,000 years ago.
2) Acupuncture points.
3) These are said to be caused by blockages on an energy pathway, or by an excess of energy.
4) Arthritis, backache, chronic headache.

REMEMBERING DETAILS

1) T
2) F Acupuncture was used only to treat various health problems by restoring a natural flow of energy in the body.
3) F In China, acupuncture anaesthesia is also used during many different kinds of medical operations, including major surgery on the brain, the kidneys, the heart, and the lungs.
4) F The key principle in acupuncture has been the treatment of the human body through its own resources, without introducing unnatural chemicals.

INFERENCES

1) c
2) c

WORD POWER

1) c
2) a
3) c
4) a
5) b
6) a

CROSSWORD PUZZLE

ACROSS: 3) addiction 5) energy 8) patients 11) circulating 13) sensation 14) interest

DOWN: 1) various 2) reduced 4) operations 6) resources 7) inserted 9) discomfort 10) dating 12) clinics

Earthquakes

1 Although earthquakes have the reputation of being very destructive, which can be true, they are in fact commonplace events. Earthquakes are usually the result of pressure that builds up below the earth's surface due to the slow but continual movement of the earth's crust. When the pressure gets too great, the great rock masses beneath the earth's surface slip and slide causing the ground above to shake.

2 Close to 6,000 earthquakes are recorded throughout the world in an average year. Of these, almost 5,500 involve earthquake shock waves that are too small to be felt. Around 450 are felt but cause no damage, and around 35 cause minor damage from rock slides and avalanches. It is the remaining 15 earthquakes per year that cause major damage and often result in people losing their lives.

3 Probably the most famous earthquake in history was the 1906 San Francisco earthquake in California. That tremor caused the destruction of 521 blocks in the city center and also started an uncontrollable fire that lasted for three days. In 1923, approximately 143,000 people died in the biggest earthquake that Tokyo, Japan ever experienced.

4 The earth's crust is divided into 15 rigid layers called plates. These plates move independently of each other and are continually either colliding or sliding past one another. The Pacific Plate is the largest in the world, stretching some 24,000 miles around the rim of the Pacific Ocean. This plate

crosses over the North American Plate in a well-known area called the San Andreas fault.

5 This famous fault line is visible above ground as a continuous fracture for over 600 miles. It stretches from San Diego, a large city in southern California, to the small town of Point Arena on the northern California coast. The disastrous San Francisco earthquake of 1906 happened on this fault line and it is feared that another major earthquake will destroy one of the big cities in California that are built along the San Andreas fault.

6 Most earthquakes can only be detected by sensitive instruments called seismographs. There are various types of seismographs that measure earthquakes differently, but the one we hear the most about is the Richter Magnitude Scale. The Richter Scale grades earthquakes on a 1-to-10 scale and is based on the amount of energy released by the rock movements. For instance, the 1995 earthquake in Kobe, Japan that killed nearly 5,500 people and left hundreds of thousands homeless registered 6.9 on the Richter Scale.

The best protection against a large, 7
destructive earthquake is early prediction of its location, time and strength. Scientists use the information they have gathered from earthquake observatories and past records of earthquakes to anticipate where a large earthquake will occur. One theory is that areas along a fault line where there has not been a recent earthquake are most likely to release built-up pressures.

Another effective method of reducing 8
the destructiveness of earthquakes is to design and build structures that will survive strong tremors. Several nations, including Japan and the United States, have created strict building codes that are legally enforced in high-risk earthquake areas. These building codes have had some success in reducing earthquake-related deaths and damage.

THE MAIN IDEA

Circle the letter of the sentence or phrase which best describes the main idea of the article about earthquakes. Be prepared to support your answer.

a) Damage caused by major earthquakes.
b) Causes and effects of earthquakes.
c) How earthquakes are detected.
d) Measures to reduce earthquake damage.

UNDERSTANDING WHAT YOU READ

If you can, answer these questions from memory. If you cannot, look back at the article.

1) About how many tremors cause major damage in an average year?

___

2) When was the San Francisco earthquake? What damage did it cause?

___

3) What is the Richter Scale?

___

4) Where do scientists think earthquakes may be most likely to happen?

___

REMEMBERING DETAILS

Write TRUE or FALSE under each statement. If the statement is false, write the statement correctly.

1) Earthquakes are quite rare events.

___

2) The Kobe earthquake was the biggest ever known in Japan.

___

3) The San Andreas fault is more than 600 miles long.

___

4) The best protection against earthquakes is strong buildings that resist tremors.

___

INFERENCES

Based on the article, circle the letter of the best sentence completion.

1) In view of how they seem to be caused, earthquakes should be...

a) easy to prevent.
b) quite hard to prevent.
c) very costly.
d) impossible to prevent.

2) The main reason why it is useful to know where and when a large earthquake is likely to happen is that...

a) people will be able to resist the quake.
b) people will be able to move to other areas.
c) people will be able to make their homes stronger.
d) people will be able to stop the quake from happening.

INTERPRETATION

1) Are earthquakes common in the area where you live, or in a place where you have lived in the past? What experience have you had with earthquakes?

2) If you were offered an excellent job in a city known for dangerous earthquakes, would you move there? Why or why not?

3) What natural events, other than earthquakes, have the power to cause great damage? Which are the worst ones? What can we do about them?

WORD POWER

Circle the letter of the word that means the same as the word on the left.

1) commonplace	a) usual	b) dangerous	c) exceptional
2) continual	a) occasional	b) constant	c) continental
3) colliding	a) joining	b) breaking	c) impacting
4) rim	a) surface	b) edge	c) bottom
5) fracture	a) break	b) valley	c) river
6) anticipate	a) protect	b) observe	c) predict

CROSSWORD PUZZLE

ACROSS:

2) The great rock masses beneath the earth's ____________________ slip and slide.

3) Several nations have created ____________________ building codes in high-risk areas.

5) Building codes have had some success in ____________________ earthquake-related damage.

8) Earthquakes have the ____________________ of being very destructive.

9) The most famous earthquake in ____________________ was the 1906 San Francisco earthquake.

11) The earth's crust is divided into 15 ____________________ layers called plates.

12) The San Francisco earthquake happened along the San Andreas ____________________.

13) There are various types of ____________________ that measure earthquakes differently.

DOWN:

1) The best protection against a large, ____________________ earthquake is early prediction.

4) Close to 6,000 earthquakes are recorded ____________________ the world in an average year.

5) The Richter Scale is based on the amount of energy ____________________ by the rock movements.

6) The tremor caused the ____________________ of 521 blocks in the city center.

7) Point Arena is located on the ____________________ California coast.

10) The Pacific Plate is the ____________________ in the world, stretching some 24,000 miles.

ANSWER KEY

THE MAIN IDEA

b) Causes and effects of earthquakes.

UNDERSTANDING WHAT YOU READ

1) In an average year, major damage is caused by about fifteen earthquakes.
2) The San Francisco earthquake occurred in 1906; 521 blocks of the city center were destroyed, and a huge fire lasted for three days.
3) The Richter Scale is a system to measure the power of earthquakes, going from a level of 1 to 10.
4) Scientists think tremors are most likely to occur along a fault line where there has been no recent earthquake.

REMEMBERING DETAILS

1) F Earthquakes are quite commonplace
2) F Japan's biggest earthquake was in Tokyo, in 1923.
3) T
4) F Strong buildings can help, but the main thing is to predict the time, place and strength of earthquakes before they happen.

INFERENCES

1) d
2) b

WORD POWER

1) a
2) b
3) c
4) b
5) a
6) c

CROSSWORD PUZZLE

ACROSS: 2) surface 3) strict 5) reducing 8) reputation 9) history 11) rigid 12) fault 13) seismographs

DOWN: 1) destructive 4) throughout 5) released 6) destruction 7) northern 10) largest

Identical Twins

1 Identical twins are brothers or sisters who look almost exactly alike. They start life from the same egg. Their genetic make-up is the same; therefore, identical twins are the same sex. They not only look the same, but sometimes they even act the same. And the fascinating effects of identical twins' genetic similarity may be expressed in many ways.

2 Just looking so much the same can have consequences for identical twins. It is easy to forget that, although twins look similar, they do not necessarily have identical needs or wishes! Yet, a tired care-giver may put twin toddlers to bed at the same time when one is sleepy, whereas the other is still alert. Or parents may dress twins exactly alike, to emphasize their strikingly similar appearance. Such actions can sometimes have the effect of undermining twins' identity as two separate individuals.

On the other hand, there have been many cases where identical twins happily shared the same experiences. Two famous American journalists, Ann Landers and Dear Abby, were twins who shared the same double bed, went to the same school, and were inseparable until the day they got married... in a double wedding ceremony! And of course they both wrote advice columns in the newspaper. These contented, well-adjusted twins' lives were identical in many respects. 3

Indeed, identical twins often display surprising psychological similarities. Two identical twins, for instance, 4 produced the same answers on an examination; so, their teacher suspected they were cheating. For the next exam, the teacher put them in different rooms. Still, they had the same mistakes and the same correct answers. This effect is not uncommon. Apart from merely looking alike, twins often tend to have similar thinking and problem-solving skills.

5 Even when they are raised in separate environments, identical twins may turn out to be remarkably similar. One case involved twin brothers who were adopted into separate families at age four months, and who never met again for 39 years. By chance, they happened to receive the same first name, Jim. But it is unlikely that chance caused other striking similarities between them. For example, they had the same hobby, carpentry, and they both built a bench around a tree in their gardens. And they even both painted it the same color, white.

6 Furthermore, both twins took holidays at the same Florida beach. Both drove the same model of car. They both had dogs with the same names. Both of them were divorced from first wives named Linda, and both married second wives named Betty. They had the same physical features - such as pulse rate, blood pressure, and weight -and they even suffered from the same kind of headache. Both chewed their fingernails, and both crossed their legs the same way.

Still, identical twins remain two separate people; they are never exactly alike. One thing or another will distinguish them: fingerprints, freckles, the way they smile, and so on. Their shared genes no doubt promote similarities in the appearance and personality of these divided eggs, but at the same time their entire life experience is equally important in causing differences between them. The effects of genetics always interact with the consequences of experience and free choice. 7

THE MAIN IDEA

Circle the letter of the sentence or phrase which best describes the main idea of the article about identical twins. Be prepared to support your answer.

a) Identical twins are very similar; so, people tend to treat them the same way.
b) Despite being very much alike in many ways, identical twins are two separate people who always have some differences.
c) Even when raised separately, identical twins have the same behavior and physical features.
d) Identical twins illustrate the fascinating effects of genetic similarity, which are studied by psychologists all over the world.

UNDERSTANDING WHAT YOU READ

If you can, answer these questions from memory. If you cannot, look back at the article.

1) Why are identical twins the same sex?

____

2) What do people often forget when dealing with identical twins?

____

3) What do we learn from the example of the identical twins who wrote the same examination answers?

____

4) What physical features did the identical twins mentioned in the story have in common?

____

REMEMBERING DETAILS

Write TRUE or FALSE under each statement. If the statement is false, write the statement correctly.

1) All twins like it when their parents dress them the same.

____

2) The teacher had no suspicions when the twins had the same answers on a test.

____

3) The twins had the same kind of dog, but the dogs had different names.

____

4) Their entire life experience is equally important in causing differences between them.

____

INFERENCES

Based on the article, circle the letter of the best sentence completion.

1) Identical twins probably share many of the same character traits because...

a) they lived together.
b) their parents dressed them the same.
c) they came from the same egg.
d) they took all the same classes.

2) It seems identical twins...

a) all have issues with their parents.
b) all wish they were not twins.
c) are famous for giving advice.
d) want to be recognized as two separate people.

INTERPRETATION

1) Do you know or have you known any identical twins? Did they seem alike?

2) Would you like to be a twin? What are the advantages and the drawbacks of being a twin?

3) Do you have brothers and sisters? Would you say the members of your family are much alike or very different from each other?

WORD POWER

Circle the letter of the word that means the same as the word on the left.

1) similarity	a) anxiety	b) alikeness	c) diversity
2) emphasize	a) stress	b) duplicate	c) moderate
3) respects	a) agreements	b) notifications	c) ways
4) merely	a) openly	b) only	c) boldly
5) chewed	a) bit	b) squeezed	c) molded
6) promote	a) help forward	b) withhold	c) attempt

CROSSWORD PUZZLE

ACROSS:

1) Two famous American ____________________ were married in a double wedding ceremony!

3) One case involved twin brothers who were ____________________ into separate families at age four months, and who never met again for 39 years.

7) Just looking so much the same can have ____________________ for identical twins.

11) They had the same physical ____________________.

12) Such actions can sometimes have the effect of undermining twins' ____________________ as two separate individuals.

13) For example, they had the same hobby, ____________________.

14) One thing or another will ____________________ them.

DOWN:

2) They start ____________________ from the same egg.

4) Identical twins often ____________________ surprising psychological similarities.

5) Furthermore, both twins took ____________________ at the same Florida beach.

6) Their teacher ____________________ they were cheating.

8) They not only look the same, but ____________________ they even act the same.

9) For the next ____________________, the teacher put them in different rooms.

10) Both drove the same ____________________ of car.

ANSWER KEY

THE MAIN IDEA

b) Despite being very much alike in many ways, identical twins are two separate people who always have some differences.

UNDERSTANDING WHAT YOU READ

1) They start life from the same egg.
2) Although identical twins look very similar, they do not necessarily have the same needs or wishes.
3) Apart from looking alike, identical twins often think very much alike, too.
4) Pulse rate, weight, and they even suffered from the same kind of headache.

REMEMBERING DETAILS

1) F Actions, such as dressing twins the same, can sometimes have the effect of undermining twins' identity as two separate individuals.
2) F Their teacher suspected they were cheating.
3) F They both had dogs with the same names. It is not mentioned if they were the same kind of dog.
4) T

INFERENCES

1) c
2) d

WORD POWER

1) b
2) a
3) c
4) b
5) a
6) a

CROSSWORD PUZZLE

ACROSS: 1) journalists 3) adopted 7) consequences 11) features 12) identity 13) carpentry 14) distinguish

DOWN: 2) life 4) display 5) holidays 6) suspected 8) sometimes 9) exam 10) model

Lasers

1 Laser is short for "light amplification by stimulated emission of radiation." Gordon Gould, a physicist, invented the laser in 1957 and the first working model was built by Theodore Maiman in 1960. It is safe to say that the laser's inventors had no idea how many uses there would be for their invention.

2 The first laser worked like this: the opposite ends of a ruby rod were polished shiny like a mirror. At one end of the rod there was a lens. The rod was full of atoms of gas. Then light from a flashing lamp entered the ruby and caused the atoms to move more rapidly. The atoms interacted with the light by bouncing light waves back and forth between the polished ends. The light bounced back and forth faster and faster until it burst through the lens as laser light. Unlike the light waves of an ordinary light, which are all different, those of a laser light are all the same wavelength.

The light beam a laser produces can 3
be focused down to a point as precise as one ten-thousandth of an inch wide. The narrower a laser beam is made, the more powerful it becomes. When used in this manner a laser beam can reach temperatures higher than 10,000 degrees Fahrenheit. And there are lasers that can produce more than 100,000 watts of energy.

Lasers have become an indispensible 4
tool in the field of medicine. Because lasers can apply energy quickly to very small areas, laser beams are used in

delicate eye surgery. To repair a detached retina, a beam of light is aimed through the pupil onto the damaged area. This action forms a tiny scar that "welds" the retina back in place. All this can be done without making a cut in the patient's eye. Lasers are used in several other medical interventions as well.

5 In the mid-1980s, US President Ronald Reagan supported the military use of lasers. Reagan promoted the building of an outer space defense system using lasers. The media nicknamed it the "Star Wars Defense" because it was supposed to be able to destroy dangerous satellites and nuclear missiles miles above the earth in outer space. The program was eventually voted down by the U.S. Congress.

6 Probably the most revolutionary application of lasers so far is in the field of communications. The high frequency of concentrated laser light makes it possible for a laser beam to carry a lot more information than radio waves or copper wire. A small number of laser beams moving along tiny glass fibers can carry about 40,000 telephone calls at once. Today many telephone calls made across oceans are carried by lasers.

Cutting diamonds and other hard metals, playing compact discs, making holograms and marking straight lines with accuracy are just a few of the other ways lasers are used. A beam of laser light was even sent from earth all the way to the moon, where it bounced off a quartz reflector (left there by astronauts) and returned to earth. 7

Lasers are used for many more things as well, too numerous to mention. But the most amazing use of lasers might only be discovered in the future. Scientists are currently trying to solve the world's energy problems by doing nuclear fusion experiments using lasers. It would not be surprising if lasers can do the job. Look how many other things they can do! 8

THE MAIN IDEA

Circle the letter of the sentence or phrase which best describes the main idea of the article about lasers. Be prepared to support your answer.

a) How lasers get their power.
b) Applications of laser beams.
c) Origins and uses of the laser.
d) Future potential of the laser.

UNDERSTANDING WHAT YOU READ

If you can, answer these questions from memory. If you cannot, look back at the article.

1) What is the difference between the light waves of an ordinary light and the light waves of a laser light?

2) How hot can laser beams become?

3) Why are lasers useful in delicate eye surgery?

4) What may be the next great application of lasers?

REMEMBERING DETAILS

Write TRUE or FALSE under each statement. If the statement is false, write the statement correctly.

1) The first actual laser device was produced in 1960.

2) The wider a laser beam is, the stronger it becomes.

3) The "Star Wars Defense" system is now in place.

4) Lasers have not yet been used for communications.

INFERENCES

Based on the article, circle the letter of the best sentence completion.

1) Laser technology is especially fascinating because...

a) its inventors never knew lasers could do so much..
b) more people are able to have eye surgery.
c) lasers have so many uses.
d) lasers can reach such high temperatures.

2) The different present-day applications of the laser all seem to rely one way on another on the basic fact that...

a) laser beams can be made to become narrow and strong.
b) laser beams can carry several messages at once.
c) laser beams begin in a ruby rod.
d) laser beams can be used in outer space.

INTERPRETATION

1) Why do you think the idea of a laser beam fascinates people so much? Are there other scientific discoveries that impress people in the same way?

2) Should scientists who find very useful processes, like laser technology, be able to sell them and become rich? Or should they work only for the good of human beings, with no business connections?

3) Can you think of other recent scientific inventions which, like the laser, have many everyday uses? Give examples and explain their importance.

WORD POWER

Circle the letter of the word that means the same as the word on the left.

1) precise	a) obvious	b) exact	c) vague
2) promoted	a) sold	b) passed	c) recommended
3) voted down	a) approved	b) studied	c) defeated
4) fibers	a) threads	b) wires	c) rods
5) accuracy	a) exactness	b) lengthiness	c) ink
6) currently	a) directly	b) presently	c) previously

CROSSWORD PUZZLE

ACROSS:

2) Laser is short for "light ____________________ by stimulated emission of radiation."

7) President Reagan promoted the building of an outer space ____________________ system.

10) There are lasers that can produce more than 100,000 ____________________ of energy.

11) The most amazing use of lasers might only be discovered in the ____________________.

13) The atoms interacted with the ____________________ by bouncing light waves back and forth.

DOWN:

1) This action forms a tiny scar that "____________________" the retina back in place.

2) The rod was full of ____________________ of gas.

3) Lasers have become an ____________________ tool in the field of medicine.

4) The laser's ____________________ had no idea how many uses there would be for their invention.

5) Lasers are used in several other medical ____________________ as well.

6) The narrower a laser ____________________ is made, the more powerful it becomes.

8) Laser beams are used in delicate ____________________ surgery.

9) Today many telephone calls made across ____________________ are carried by lasers.

12) The opposite ends of a ruby ____________________ were polished shiny like a mirror.

ANSWER KEY

THE MAIN IDEA

c) Origin and uses of the laser.

UNDERSTANDING WHAT YOU READ

1) The difference between the light waves of an ordinary light and the light waves of a laser light is that the light waves of an ordinary light are all different wavelengths, while those of a laser light are all the same wavelength.
2) Lasers can reach temperatures higher than 10,000 degrees Fahrenheit.
3) Lasers are useful in delicate eye surgery because they can apply energy quickly to a very small area.
4) The next great application of lasers may be to solve the world's energy problems.

REMEMBERING DETAILS

1) T
2) F The narrower a laser beam is made, the more powerful it becomes.
3) F The "Star Wars Defense" system is not now in place. The program was eventually voted down by the U.S. Congress.
4) F Lasers are used a lot in communications because high frequency laser light moving along a glass fiber can carry many messages at once.

INFERENCES

1) c
2) a

WORD POWER

1) b
2) c
3) c
4) a
5) a
6) b

CROSSWORD PUZZLE

ACROSS: 2) amplification 7) defense 10) watts 11) future 13) light

DOWN: 1) welds 2) atoms 3) indispensible 4) inventors 5) interventions 6) beam 8) eye 9) oceans 12) rod

Astrology

1 Since the first day turned into night, people have looked upward and wondered what effect, if any, the moon, the stars and the other planets have on them. The observation and charting of the movements of certain heavenly bodies has became known as astrology. Today astrology is used to make predictions about events and about people's futures.

2 The use of astrology to predict human events originated in ancient Babylon and spread to China, India and the West. By the 12th century astrology was considered to be a reliable science. It prospered and remained popular until the 17th century when Galileo (inventor of the telescope) discovered that the sun, not the earth, was the center of our galaxy. After that change in perspective, astrology was generally considered to be nothing more than superstition.

Most of the predictions derived 3
from astrology revolve around a person's horoscope. Astrologers divide the year into 12 sections, called the zodiac. The 12 signs of the zodiac match the 12 constellations (groups of stars) lying in a great belt around our galaxy. One's horoscope is a map of the night sky at the time of one's birth. A person's date of birth determines his or her zodiac sign. For example, a person born between July 23 and August 22 has the sign of Leo because of the position of the Leo constellation between those dates.

4 Based on zodiac signs, people can check their horoscope for the general outlook of their day, week or long-term future. Or a person could have a personal horoscope drawn up for more specific information. Some people regularly consult an astrologer about the most favorable times and dates for major events in their lives. Other people are known to use astrological information and horoscopes to pick lottery numbers and choose potential spouses or careers.

5 People often confuse astrology with astronomy. This is partly because at one time the meanings of the two words were almost interchangeable. Today astronomy is still the study of the stars and planets; however, unlike astrology, astronomy is an exact science. The astronomer Galileo and his telescope helped reveal many facts about the heavens that were previously unknown. Over the centuries, the use of more sophisticated equipment has helped astronomers and scientists discover what stars are made of, the existence of other galaxies, and the origins and future of our solar system.

For hundreds of years critics have 6
been questioning the validity of astrology. Fact-based scientists have always denied the power of celestial activity either to influence people and events or to predict the future. They lump astrology in with palm-reading, numerology, and crystal ball gazing, as non-scientific and lacking objective evidence.

But despite the challenges, astrology 7
has survived to the present day. Currently, a person can check any local or national newspaper for his or her horoscope. In North America in addition to newspapers, some psychics have television shows and others have telephone numbers which offer people a chance to talk about their futures...for a fee! Even in today's modern scientific era, many people are still looking for answers to their questions about an uncertain and often frightening future.

THE MAIN IDEA

Circle the letter of the sentence or phrase which best describes the main idea of the article about astrology. Be prepared to support your answer.

a) When did astrology begin?
b) What is astrology?
c) How does astrology work?
d) Why do people rely on astrology?

UNDERSTANDING WHAT YOU READ

If you can, answer these questions from memory. If you cannot, look back at the article.

1) How do astrologers attempt to gather information?

____

2) Where did astrology begin and develop, before it reached the West?

____

3) Where can people find their horoscope?

____

4) What do astronomers do?

____

REMEMBERING DETAILS

Write TRUE or FALSE under each statement. If the statement is false, write the statement correctly.

1) Galileo was an important astrologer.

____

2) In astrology, a person's sign is determined by an animal that was seen at the time of birth.

____

3) Only recently have scientists been in a position to criticize astrology.

____

4) In an uncertain and frightening world, some people need the comfort brought by astrology.

____

INFERENCES

Based on the article, circle the letter of the best sentence completion.

1) People who use astrology to help them make important decisions seem to be...

a) following tradition rather than logic.
b) doing it just for fun.
c) seeing the future very clearly.
d) becoming very rich and successful.

2) Scientists reject the claims of astrology because...

a) they do not want to share their power with astrologers.
b) they say there is no solid proof to support astrology's claims.
c) they believe that astrology is old-fashioned and out-of-date.
d) they accept astrology itself, but not related theories like palm-reading, numerology, and crystal ball gazing.

INTERPRETATION

1) What is your zodiac sign? Do you think your personality matches your sign? Give examples.

2) Have you ever read a newspaper horoscope that really did predict something that was going to happen, or that helped you make an important decision? Explain.

3) For you, are horoscopes worth considering seriously?... or just a game?...or a bit of each? Explain your reasons.

WORD POWER

Circle the letter of the word that means the same as the word on the left.

1) originated	a) continued	b) ended	c) began
2) reliable	a) trustworthy	b) important	c) popular
3) derived from	a) spread from	b) obtained from	c) prevented from
4) spouses	a) business partners	b) tennis partners	c) marriage partners
5) lump in	a) criticize	b) include	c) ridicule
6) era	a) time	b) knowledge	c) theory

CROSSWORD PUZZLE

ACROSS:

4) After Galileo's discovery, astrology was generally considered to be ____________________.

8) Unlike astrology, astronomy is an ____________________ science.

9) A person's date of birth determines his or her zodiac ____________________.

10) Galileo discovered that the sun, not the earth, was the center of our ____________________.

12) Predictions derived from astrology revolve around a person's ____________________.

13) People often confuse astrology with ____________________.

14) For hundreds of years critics have been questioning the ____________________ of astrology.

DOWN:

1) Today ____________________ is used to make predictions about the future of people's lives.

2) Astrology is the observation and charting of the movements of ____________________ bodies.

3) One's horoscope is a map of the night sky at the time of one's ____________________.

5) By the 12th century astrology was considered to be a reliable ____________________.

6) At one time the meanings of the two words were almost ____________________.

7) People can check any local or national ____________________ for their horoscope.

11) Astrologers divide the year into 12 sections, called the ____________________.

ANSWER KEY

THE MAIN IDEA

b) What is astrology?

UNDERSTANDING WHAT YOU READ

1) Astrologers attempt to gather information by observing and charting the movements of heavenly bodies.
2) Astrology began in ancient Babylon, and then it spread to China and India.
3) People can find their horoscope in the newspaper or, if they want more detailed information, they can have a personal horoscope drawn up. In addition, there are telephone horoscope services.
4) Like astrologers, astronomers study the stars and planets, but their goal is different: they want to learn about the solar system.

REMEMBERING DETAILS

1) F Galileo's discovery that the sun, not the earth, was the center of the galaxy went against astrology.
2) F A person's sign is determined by the date of his or her birth, related to one of twelve constellations of stars associated with the time of the year.
3) F In fact, scientists have been attacking astrology for hundreds of years.
4) T

INFERENCES

1) a
2) b

WORD POWER

1) c
2) a
3) b
4) c
5) b
6) a

CROSSWORD PUZZLE

ACROSS: 4) superstition 8) exact 9) sign 10) galaxy 12) horoscope 13) astronomy 14) validity

DOWN: 1) astrology 2) heavenly 3) birth 5) science 6) interchangeable 7) newspaper 11) zodiac

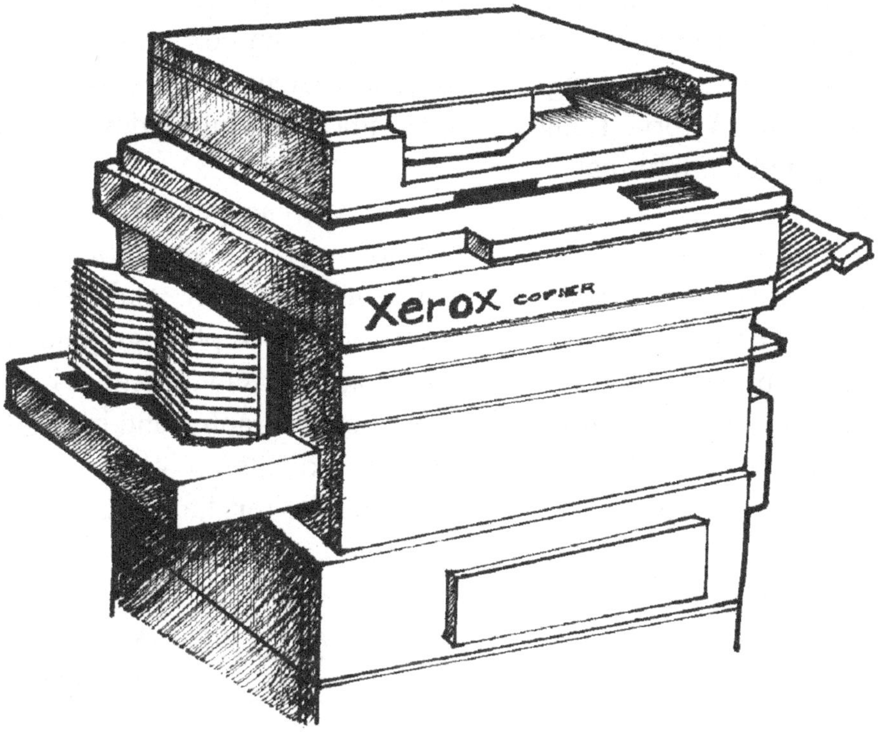

Xerography

1 What do you find in many offices, schools, libraries and hospitals, and even in many homes? Photocopiers. Photocopiers are everywhere. The story behind the invention of the photocopier began in 1937 when a thirty-one year old named Chester Carlson applied for a patent on his invention, a wooden box for making copies. Today that wooden box sits in the Smithsonian Institution's Hall of Photography. The process by which Carlson's machine made copies later came to be known as xerography.

2 Chester Carlson was born in Seattle in 1906. His parents were very poor, and his father suffered from severe arthritis. Chet had to work at odd jobs to support his parents because his father could not work. When his mother died, Chet was left to support and care for his incapacitated father. Chet was seventeen. In spite of this hardship, he worked his way through college and graduated with a degree in physics at age twenty-four.

With his degree and a keen interest in 3
providing the world with a better copying method, Carlson got to work reading every book he could find that had anything to do with printing, photocopying and copying. Then he began experimenting. Success did not come easily or quickly. But although his process of "electrophotography" was little more than an idea, in 1937 Chester Carlson applied for a patent for his invention.

4 October 22, 1938 is a very important date in the history of xerography. That was the date when Carlson, working with an assistant in his Astoria, Long Island laboratory, had his first real success. He managed to copy 10-22-38 Astoria, the date and the place of the experiment. Carlson was ecstatic. He had his first real proof that electrophotography might actually work. Today, there is a plaque on the house marking the spot where Chester Carlson invented xerography.

5 One of the problems Carlson faced was a lack of money to continue his experiments. He tried to interest some large companies in his idea, but most were not interested. And those that allowed Carlson to do demonstrations were turned off because the copied images were smudged or the paper was discolored. These rejections were difficult for Chet to accept. Fortunately, he did not give up.

6 Finally, someone showed an interest. The Batelle Memorial Institute of Columbus, Ohio was a privately funded research institute. The Institute's engineers saw potential in Chet's wooden box. Along with Chet, Batelle's scientists did countless experiments trying to develop a copier.

In 1945, managers of the Haloid Company read a description of Carlson's invention. Haloid was a manufacturer of photographic paper. The company saw potential in electrophotography, and by 1946 Haloid had entered into an agreement with Batelle for the rights to Chet's invention. 7

Eventually, Carlson's process was perfected. But the company needed a name for the copying machine. To them the name electrophotography was not catchy enough. They settled on xerography. "Xero" was the Greek word for "dry" and "graphein" meant writing or drawings. The name fit perfectly because that is what the machine did: it made dry copies. So they called the machine a Xerox machine and soon Haloid changed its name to Xerox as well. And the rest is history. 8

THE MAIN IDEA

Circle the letter of the sentence or phrase which best describes the main idea of the article about xerography. Be prepared to support your answer.

a) Chester Carlson's struggle to make the photocopier a success.
b) How Chester Carlson got the idea for the photocopier.
c) People who helped Chester Carlson develop the photocopier.
d) Chester Carlson's work with Xerox, the giant photocopier company.

UNDERSTANDING WHAT YOU READ

If you can, answer these questions from memory. If you cannot, look back at the article.

1) Describe Carlson's childhood.

2) What happened on October 22, 1938? Why was it important?

3) What delayed Carlson's work? How did he finally solve this problem?

4) How did Haloid help in developing the photocopier?

REMEMBERING DETAILS

Write TRUE or FALSE under each statement. If the statement is false, write the statement correctly.

1) Because he was so poor, Carlson could not study for a university degree.

2) Carlson applied for a patent as soon as he had perfected a functioning photocopier.

3) Many possible investors refused to give Carlson money for his work.

4) Carlson eventually founded the Xerox Company.

INFERENCES

Based on the article, circle the letter of the best sentence completion.

1) At first people thought that...

a) Carlson would take a long time to perfect his photocopier.
b) Carlson's photocopier would be too expensive to be practical.
c) Carlson did not want to let others invest in his photocopier.
d) Carlson's idea for a photocopier would never come to anything.

2) One of the most surprising elements in this story is the fact that...

a) a common machine like the photocopier, which we take for granted, took so long to develop.
b) the photocopier, a very ordinary machine, was invented by a brilliant scientist.
c) a man was willing to devote his whole life to a boring machine like the photocopier.
d) the Batelle Memorial Institute took an interest in the photocopier.

INTERPRETATION

1) Do you think the photocopier is a good machine or a bad one? What are its positive and negative features?

2) What personal qualities did Chester Carlson need in order to succeed? Can you think of other people whose inventions have changed your life?

3) Many people use photocopiers all the time without thinking about it. How many times in the last week have you benefited from the invention of xerography? Explain your answer.

WORD POWER

Circle the letter of the word that means the same as the word on the left.

1) hardship	a) hard work	b) bad trouble	c) difficult subject
2) assistant	a) helper	b) worker	c) engineer
3) ecstatic	a) happy	b) delighted	c) pleased
4) smudged	a) backwards	b) small	c) unclear
5) funded	a) furnished	b) founded	c) financed
6) catchy	a) memorable	b) beautiful	c) modern

CROSSWORD PUZZLE

ACROSS:

1) On Oct. 22, 1938, Carlson had his first ____________________ electrophotography might actually work.

4) Success for Carlson did not come quickly, nor ____________________.

7) The Institute's engineers saw potential in Chet's ____________________ box.

8) Along with Chet, Batelle's scientists did ____________________ experiments.

10) The process by which the machine made ____________________ came to be known as xerography.

12) Today, there is a ____________________ on the house where Chester Carlson invented xerography.

13) Eventually, Carlson's ____________________ was perfected.

14) In 1945, managers of the Haloid Company read a description of Carlson's____________________.

15) Carlson's father ____________________ from severe arthritis.

DOWN:

1) Carlson applied for a ____________________ on his invention.

2) Chet had to work ____________________ jobs to support his parents.

3) One of the problems Carlson faced was a lack of ____________________ to continue his experiments.

5) Some were turned off because the copied ____________________ were smudged.

6) Chet was left to support and care for his ____________________ father.

9) Carlson got to work ____________________ anything he could find having to do with copying.

11) Carlson managed to copy the date and the ____________________ of the experiment.

ANSWER KEY

THE MAIN IDEA

a) Chester Carlson's struggle to make the photocopier a success.

UNDERSTANDING WHAT YOU READ

1) Chester Carlson was born in Seattle in 1906. His parents were very poor, and his father suffered from severe arthritis. Chet had to work at odd jobs to support his parents because his father could not work. When his mother died, Chet was left to support and care for his incapacitated father. Chet was seventeen. In spite of this hardship Chet worked his way through college and graduated with a degree in physics at age twenty-four.
2) On October 22, 1938 Chet Carlson, working with an assistant in his Astoria, Long Island laboratory, had his first real success. He managed to copy 10-22-38 Astoria, the date and the place of the experiment. Carlson had his first real proof that electrophotography might actually work.
3) Carlson's work was delayed by a lack of money to continue his experiments. He tried to interest some large companies in his idea, but most were not interested. Finally, The Batelle Memorial Institute, a privately funded research institute, saw potential in Chet's wooden box.
4) Haloid saw potential in electrophotography. By 1946 Haloid had entered into an agreement with Batelle for the rights to Chet's invention.

REMEMBERING DETAILS

1) F In spite of being poor Chet worked his way through college and graduated with a degree in physics at age twenty-four.
2) F Chester Carlson applied for a patent for his invention even though his process of "electrophotography" was little more than an idea.
3) T
4) F The Xerox Company was formerly known as Haloid. Haloid entered into an agreement with Batelle for the rights to Chet's invention. Chet Carlson did not found the Xerox Company, nor did he found Haloid. These companies just used his invention.

INFERENCES

1) d
2) a

WORD POWER

1) b
2) a
3) b
4) c
5) c
6) a

CROSSWORD PUZZLE

ACROSS: 1) proof 4) easily 7) wooden 8) countless 10) copies 12) plaque 13) process 14) invention 15) suffered

DOWN: 1) patent 2) odd 3) money 5) images 6) incapacitated 9) reading 11) place

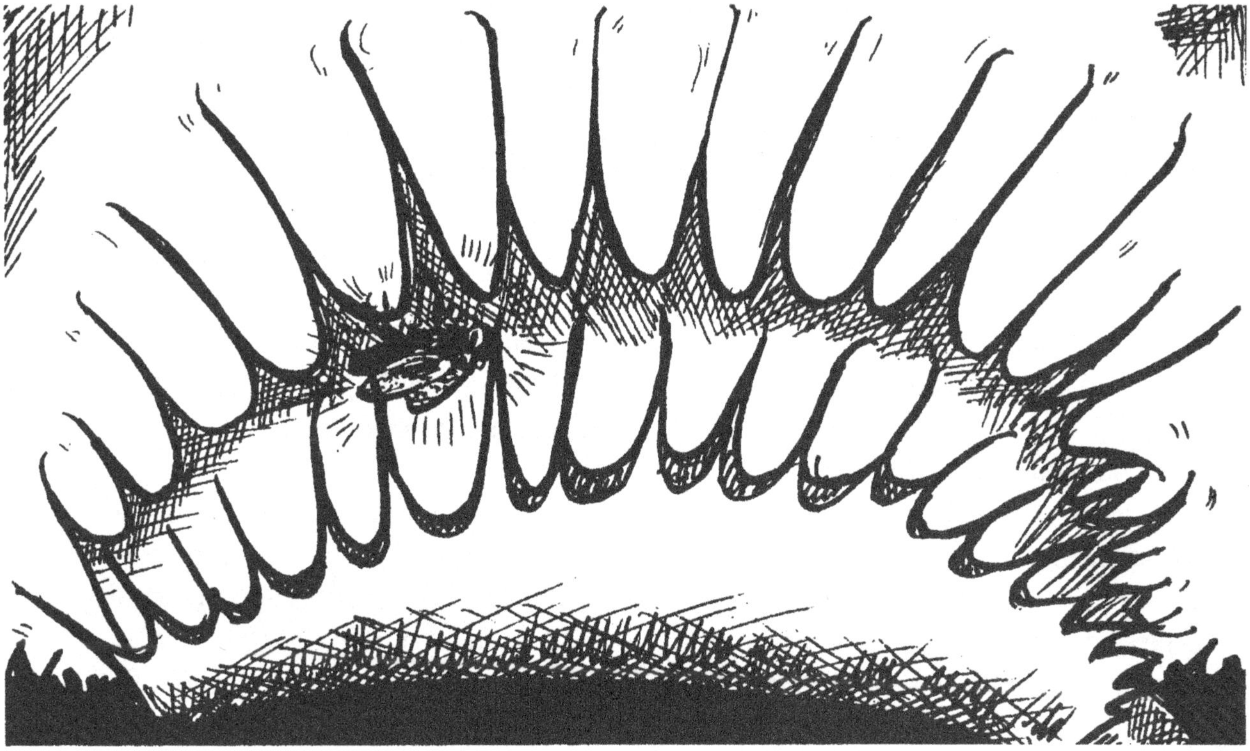

Carnivorous Plants

1 On a warm sunny morning, a tiny bee hovers over a bright red flower that is swaying in the gentle breeze. After a moment's hesitation, the bee finds the blossom too attractive to resist. It lands on the lovely flower. Within seconds, the crimson petals close; they are edged with tightly interlocking spines that permit no escape. And that's the end of our bee!

2 The bee has been eaten by a carnivorous plant - a plant that eats living, breathing creatures, mostly creeping insects or small flies. Sometimes we call such plants insectivorous because that is what they eat. These hunters generally do not look as scary as their behavior might suggest. Most do not even have strong, woody stems. The majority of carnivorous plants are delicate climbers with beautiful, bright-colored leaves and flowers.

All carnivorous plants are perennial; in 3
other words, they continue living from year to year. They are not hard to please: in fact, they can take any climate - from arctic to tropical, and from sea level to high in the mountains. These sharks of the plant world grow on every continent, although most seem to be found in temperate, warm or tropical areas. Some adventurous plant-lovers like to keep them indoors, and they do fine in greenhouse conditions, too.

Despite these general similarities, 4
insectivorous plants do differ a lot in structure. For instance, some have pitfall traps that unwary insects may fall into;

others use sticky traps like fly-paper; and still others set mechanical traps that close on their victims. The first kind are commonly known as pitcher plants because they have a deep, liquid-filled cup a bit like a pitcher. They come in bright yellow, red, or plain green. The fluid in the pitcher smells sweet; as a result, it attracts insects, which fall down inside and are digested.

5 The best-known example of the fly paper group is sundew. It has leaves that look like flowers with striking colors: generally, pink, yellow, or white. On the surface of each cluster of leaves, sticky droplets shine like dew in the sunlight. These glistening droplets are produced by numerous tiny hairs, which cover each leaf. The droplets look like a delicious meal. But in fact, when an unlucky fly lands on them, it gets stuck and is eventually digested itself.

6 A good illustration of a carnivorous plant with a mechanical trap is Venus' flytrap, which is usually bright crimson in color. This trap really is a flower, with two roundish blades that can open wide. When an insect touches the colored surface, the blades suddenly close. Stiff spines along the edge of the blades mesh together, to keep the victim inside. Although this trickster is called Venus' flytrap, it usually preys on crawling insects such as ants.

7 Despite their impressive traps, these predators never live only on insects. They catch their prey just as supplementary nourishment. Often they grow in environments where the supply of ordinary food from the soil is limited. So, in order to thrive they have developed their remarkably specialized leaves or flowers. There is little chance that we could use these plants to rid our homes or gardens of pesky insects but, in their own terms, they are very effective hunters indeed.

THE MAIN IDEA

Circle the letter of the sentence or phrase which best describes the main idea of the article about carnivorous plants. Be prepared to support your answer.

a) Where carnivorous plants can be found.
b) The main types of carnivorous plants, and their features.
c) How appearances can deceive.
d) Carnivorous plants' use of supplementary nourishment.

UNDERSTANDING WHAT YOU READ

If you can, answer these questions from memory. If you cannot, look back at the article.

1) What is a carnivorous plant?

---

2) Briefly describe each of the main groups of carnivorous plants.

---

3) What is the use of the liquid in the pitcher plant's "pitcher"?

---

4) Why are carnivorous plants unlikely to be used by people to control insects in their homes?

REMEMBERING DETAILS

Write TRUE or FALSE under each statement. If the statement is false, write the statement correctly.

1) All carnivorous plants are annuals, which means they die after one year.

---

2) Despite these general similarities, insectivorous plants do differ a lot in structure.

---

3) The best-known example of the fly paper group is the pitcher plant.

---

4) Soft spines along the center of the blades mesh together, to keep the victim outside.

INFERENCES

Based on the article, circle the letter of the best sentence completion.

1) The Venus' flytrap feeds mostly on ants probably because...

a) flies have better eyes than ants.
b) more ants crawl inside the trap than flies land inside the trap.
c) ants have more flavor and nutrients.
d) flies are bigger than ants and they are harder for the plant to digest.

2) Carnivorous plants...

a) eat insects because insects try to steal the plant's nectar.
b) are the answer to getting rid of insects in our homes.
c) are perennials because they live in poor soil conditions.
d) eat insects to supplement their diets because they live in poor soil conditions.

INTERPRETATION

1) How do you explain the existence of carnivorous plants? Is it weird that there are plants that eat meat?

2) What is your favorite plant? Explain your answer.

3) What is the most important plant on our planet? Explain your answer.

WORD POWER

Circle the letter of the word that means the same as the word on the left.

1) resist	a) annoy	b) travel	c) say no to
2) unwary	a) unconcerned	b) aware	c) denounce
3) cluster	a) bottle	b) group	c) merge
4) crimson	a) purple	b) red	c) black
5) mesh	a) web	b) bow	c) stern
6) pesky	a) itchy	b) troublesome	c) friendly

CROSSWORD PUZZLE

ACROSS:

1) A tiny bee ____________________ over a bright red flower that is swaying in the gentle breeze.

4) These hunters generally do not look as scary as their ____________________ might suggest.

6) When an insect touches the colored surface, the blades ____________________ close.

7) Despite their impressive traps, these ____________________ never live only on insects.

9) The best-known example of the fly paper group is ____________________.

11) They are edged with ____________________ interlocking spines that permit no escape.

12) There is little chance that we could use these plants to rid our ____________________ or gardens of pesky insects.

DOWN:

2) Often they grow in environments where the supply of ordinary food from the ____________________ is limited.

3) Pitcher plants come in bright yellow, red, or plain ____________________.

4) A carnivorous plant is a plant that eats living, ____________________ creatures, mostly creeping insects or small flies.

5) These glistening droplets are produced by ____________________ tiny hairs.

7) All carnivorous plants are ____________________.

8) When an unlucky fly lands on them, it gets ____________________ and is eventually digested.

10) Despite general similarities, insectivorous plants do ____________________ a lot in structure.

ANSWER KEY

THE MAIN IDEA

b) The main types of carnivorous plants, and their features.

UNDERSTANDING WHAT YOU READ

1) A plant that eats creeping insects and small flies.
2) Plants with sticky traps; plants with pitfall traps; and plants with mechanical traps.
3) It attracts insects to its sweet smell.
4) They do not catch enough insects for this to be worthwhile.

REMEMBERING DETAILS

1) F All carnivorous plants are perennial; in other words, they continue living from year to year.
2) T
3) F The best-known example of the fly paper group is the sundew.
4) F Stiff spines along the edge of the blades mesh together, to keep the victim inside.

INFERENCES

1) b
2) d

WORD POWER

1) c
2) a
3) b
4) b
5) a
6) b

CROSSWORD PUZZLE

ACROSS: 1) hovers 4) behavior 6) suddenly 7) predators 9) sundew 11) tightly 12) homes

DOWN: 2) soil 3) green 4) breathing 5) numerous 7) perennial 8) stuck 10) differ

Automation

1 A jet pilot on a long flight over the Atlantic Ocean presses a button, and in just a few seconds a computer takes control of the flight and steers the jet. A man walks into his bank, and by using his plastic credit card he receives money from the bank machine. A woman in a factory watches over a robot as it puts together a new car piece by piece. A child sits down in front of a television, and with a quick press of the remote control, she is viewing her favorite movie.

2 In the modern world things like this happen every day. Many of the simple tasks people once had to do themselves are now being done by machines. This has made life much easier for humans by reducing the amount of work they have to do. At work, at home, in offices and factories, the impact of "automation" can be seen in almost every aspect of life.

3 "Automation" is a word that describes the way machines are used to perform the jobs that humans once did. When something becomes "automated," it continues to work even though humans are no longer involved in the process. Automation can be as simple as cooking something in a microwave oven or as complex as guiding a satellite through the unexplored regions of space.

4 For thousands of years, humans could rely only on themselves. In order to survive, they had to hunt for their own food, make their own

clothing and build their own shelter. Most of this work was done by hand, and the work was very hard. As civilization grew, humans began to make many different tools to make their lives much easier: shovels for digging, hammers for building, ovens for cooking, and wheels for traveling great distances.

5 The invention of machines and engines helped increase the speed of work around the world. Jobs that once took days for humans to finish were now being done in hours by machines.

6 In the 20th century, automation occurred faster than at any time in history. In that period, humans went from using horses for travel to flying to the moon. As we begin the new century, computers are proving to be one of the greatest inventions in history.

7 Although most people would agree that automation has improved the quality of human life, some people believe it has caused an equal number of problems. As automation occurs, many jobs humans once did begin to disappear. Machines that save companies time and money often put people out of work.

8 In huge factories, for example, automated robots are slowly taking over the job of building other products. At a car company every automated robot used on the assembly line has resulted in an average of two people losing their jobs. Without jobs, people have no money to buy products, and the entire economy suffers as a result.

9 There is no denying that automation has made life easier for humans. Machines can now do many of the dangerous jobs humans once did. Computers are used in hospitals, in food processing plants, in banks and schools. By helping us complete our work, automation can give us the freedom to enjoy the good things in life!

THE MAIN IDEA

Circle the letter of the sentence or phrase which best describes the main idea of the article about automation. Be prepared to support your answer.

a) Automation has changed the way humans live.
b) Automation has just as many good results as bad ones.
c) The modern world is full of automation.
d) The world is totally automated.

UNDERSTANDING WHAT YOU READ

If you can, answer these questions from memory. If you cannot, look back at the article.

1) How has automation made life easier for humans?

____

2) What does it mean when something becomes "automated"?

____

3) How did humans do things before so many things became automated?

____

4) Why do some people believe automation has caused problems?

____

REMEMBERING DETAILS

Write TRUE or FALSE under each statement. If the statement is false, write the statement correctly.

1) Machines have made life much more difficult for most people.

____

2) Although machines have taken jobs from humans, machines do not do the work any more quickly.

____

3) In huge factories, automated robots are slowly taking over the job of building other products.

____

4) Although machines are used for several jobs, the most dangerous work must still be performed by humans.

____

INFERENCES

Based on the article, circle the letter of the best sentence completion.

1) Before automation...

a) life was very simple.
b) everyone had a job.
c) people had tools to do their work.
d) people had to work hard to produce everything themselves.

2) When people lose their jobs because of automation...

a) the economy suffers.
b) new jobs are created for machine operators.
c) it slows the production of new products.
d) life gets easier for everyone.

INTERPRETATION

1) What automated products do you use in your everyday life? Do they make your life easier? Why or why not?

2) What would life be like if machines and automation did not exist? Would life be better or worse?

3) What types of automated products do you think will be available in the future? What types of automated products would you like to see invented? What jobs would they perform?

WORD POWER

Circle the letter of the word that means the same as the word on the left.

1) steers	a) directs	b) builds	c) engineers
2) remote	a) distant	b) automatic	c) computer
3) aspect	a) age	b) kind	c) area
4) rely on	a) trust in	b) depend on	c) understand
5) occurred	a) increased	b) developed	c) happened
6) assembly	a) fabricating	b) working	c) jiggling

CROSSWORD PUZZLE

ACROSS:

1) Tasks people once had to do themselves are now being done by ____________________.

4) "____________________" describes the way machines do the jobs that humans once did.

5) Computers are proving to be one of the ____________________ inventions in history.

6) Without jobs people have no money to buy ____________________.

7) The invention of machines helped ____________________ the speed of work around the world.

11) Automation can be as simple as cooking something in a ____________________ oven.

13) Humans have gone from using horses for travel to ____________________ to the moon.

14) People had to hunt their own food and build their own ____________________.

DOWN:

2) Most of this work was done by ____________________, and the work was very hard.

3) Machines can now do many of the ____________________ jobs humans once did.

8) A jet pilot presses a button, and a ____________________ takes control of the jet.

9) A woman watches over a robot as it puts ____________________ a new car piece by piece.

10) There is no ____________________ automation has made life easier for humans.

12) Automation has made life much ____________________ for humans.

ANSWER KEY

THE MAIN IDEA

a) Automation has changed the way humans live.

UNDERSTANDING WHAT YOU READ

1) Automation has made life much easier for humans by reducing the amount of work they have to do.
2) When something becomes "automated," it continues to work even though humans are no longer involved in the process.
3) Before so many things became automated the only thing humans could rely on was themselves. In order to survive, they had to hunt their own food, make their own clothing and build their own shelter. Most of this work was done by hand, and the work was very hard.
4) Some people believe automation has caused problems because many jobs humans once did disappeared. Machines that save companies time and money often put people out of work.

REMEMBERING DETAILS

1) F Machines have made life much easier for most humans by reducing the amount of work they have to do.
2) F Jobs that once took days for humans to finish are done in hours by machines.
3) T
4) F Machines now do many of the dangerous jobs humans once did.

INFERENCES

1) d
2) a

WORD POWER

1) a
2) a
3) c
4) b
5) c
6) a

CROSSWORD PUZZLE

ACROSS: 1) machines 4) automation 5) greatest 6) products 7) quicken 11) microwave 13) flying 14) shelter

DOWN: 2) hand 3) dangerous 8) computer 9) together 10) denying 12) easier

Lunar Landing

1 Human space travel and landing on other planets were only possible in science fiction books and movies as recently as the 1950s. But when President Kennedy committed the United States to landing a person on the moon before the end of the sixties, fiction was soon to become reality.

2 During the 1960s several manned space flights were successful and each made a contribution to furthering the distance of the next. It all started with the Mercury program, which was the earliest U.S. project to put an astronaut into space. The Mercury flights established short orbits around the earth. Next came the Gemini program, which carried two astronauts each mission. Gemini was followed by the Apollo program. Gemini and Apollo flights extended those orbital lengths until that amazing moment in history: the Apollo 11 flight.

3 It takes a lot of power to send a spacecraft to the moon. To reach the moon the Apollo 11 spacecraft needed to attain a speed of 24,300 miles per hour. This was accomplished with the power of the Saturn V rocket. Designed in three stages, the Saturn V rocket used stage one for lift-off, stage two for breaking clear of earth's gravity, and stage three for accelerating towards the moon.

4 The Apollo 11 spacecraft named Columbia lifted off at 9:32 a.m. on July 16, 1969 with a crew of Neil Armstrong, Michael Collins and Edwin "Buzz" Aldrin. The historic launch

was watched by approximately one million spectators at Cape Kennedy, Florida and by hundreds of millions around the world on television.

5 Each stage of the Saturn V rocket ignited on time and did its job. Within three days the Apollo 11 command module and the astronauts were in orbit around the moon at an altitude of about 75 miles from the surface. After the on-board systems were completely checked over, it was time to begin the important and dangerous descent to land on the moon.

6 Astronauts Armstrong and Aldrin climbed into the smaller lunar module named Eagle, which was designed for the actual moon landing. A computer controlled Eagle's descent from orbit until it was about 450 feet from the surface; then the astronauts completed the landing manually. At 4:17 p.m. on July 20, 1969 humans were on the moon.

7 It took three hours for the excited Armstrong and Aldrin to get their spacesuits on, which they had to do before they could leave the lunar module. It took another three and a half hours to depressurize Eagle's cabin. Finally, with a world-wide television audience anxiously waiting, Armstrong descended Eagle's nine-rung ladder and became the first person to walk on the moon.

8 Aldrin joined him shortly afterwards and the two men walked cautiously around and performed many experiments. It took them a while to get used to moving around because the moon has no gravity. While on the lunar surface they collected about 48 pounds of moon rock, so that it could be studied by scientists back on earth.

9 When their work was completed, Armstrong and Aldrin lifted off from the moon in Eagle and joined Collins in the orbiting Columbia. Using smaller rockets they began their return trip, landing in the Pacific Ocean on July 24, 1969. The mission was a complete success, and President Kennedy's goal had been reached.

THE MAIN IDEA

Circle the letter of the sentence or phrase which best describes the main idea of the article about the lunar landing. Be prepared to support your answer.

a) President Kennedy's contribution to the space program.
b) How a human being finally landed on the moon.
c) Space technology in the U.S.A.
d) The first activities on the lunar surface.

UNDERSTANDING WHAT YOU READ

If you can, answer these questions from memory. If you cannot, look back at the article.

1) Who was the first person to walk on the moon?

---

2) How did each successive space flight contribute to landing a person on the moon?

---

3) How could Apollo 11 attain a speed of 24,300 miles per hour?

---

4) What did the astronauts do while they were outside on the lunar surface?

REMEMBERING DETAILS

Write TRUE or FALSE under each statement. If the statement is false, write the statement correctly.

1) Hundreds of millions of people around the world watched the Apollo 11 launch because it was the first time that human beings would step on the moon.

---

2) The U.S. space program started with the Gemini program.

---

3) Apollo 11 took off from the Pacific Ocean.

---

4) Two astronauts collected moon rock, which they then studied in their lunar module.

INFERENCES

Based on the article, circle the letter of the best sentence completion.

1) Earlier missions before Apollo 11 did not go to the moon because...

a) they were unmanned spacecraft.
b) they carried only two astronauts.
c) they could only stay in orbit for a short period of time.
d) they were not strong enough to break free of earth's gravity.

2) It was difficult for the first two astronauts to walk on the moon because...

a) on earth, humans do not have to move around in zero gravity.
b) there were many people watching via television, and so the astronauts got nervous.
c) it took hours to put on their spacesuits and depressurize the cabin.
d) they had to perform many experiments while on the moon.

INTERPRETATION

1) Why did the U.S. put so much money into landing astronauts on the moon? Do you think this was a worthwhile project or would you have spent the money for another purpose?

2) What type of personality and what skills does someone need to become an astronaut?

3) Would you like to go to the moon? Would you like to live there? Why or why not?

WORD POWER

Circle the letter of the word that means the same as the word on the left.

1) mission	a) contribution	b) distance	c) flight
2) extend	a) further	b) succeed	c) establish
3) attain	a) descend	b) break	c) reach
4) accelerating	a) orbiting	b) speeding up	c) completing
5) crew	a) scientist	b) team	c) audience
6) lifted off	a) launched	b) landed	c) orbited

CROSSWORD PUZZLE

ACROSS:

1) President Kennedy committed the United States to ____________________ a person on the moon.
4) The Gemini program carried two ____________________ each mission.
6) At 4:17 p.m. on July 20, 1969 humans were on the ____________________.
7) It takes a lot of ____________________ to send a spacecraft to the moon.
9) It was time to begin the important and dangerous ____________________ to the surface of the moon.
11) The Mercury program was the ____________________ U.S. project to put an astronaut into space.
12) Within three days the Apollo 11 command ____________________ was in orbit around the moon.
14) It took three hours for the excited Armstrong and Aldrin to get their ____________________ on.
15) Armstrong descended Eagle's nine-rung ____________________ and was the first person on the moon.

DOWN:

2) Each stage of the Saturn V rocket ____________________ on time and did its job.
3) Moon rock was collected to be studied by ____________________ back on earth.
5) The mission was a complete ____________________ and President Kennedy's goal had been reached.
6) During the 1960s several ____________________ space flights were successful.
8) The Mercury flights established short ____________________ around the earth.
10) Aldrin and Armstrong walked ____________________ around and performed many experiments.
13) The smaller ____________________ module named Eagle was designed for the moon landing.

ANSWER KEY

THE MAIN IDEA

b) How a human being was finally landed on the moon.

UNDERSTANDING WHAT YOU READ

1) The first person to walk on the moon was Neil Armstrong.
2) Each successive space flight contributed to landing a person on the moon by furthering the distance of the next flight.
3) Apollo 11 could attain a speed of 24,300 miles per hour by the use of the Saturn V rocket. The Saturn V rocket used stage one for lift-off, stage two for breaking clear of earth's gravity and stage three for accelerating towards the moon.
4) On the lunar surface the two men walked cautiously around and performed many experiments. It took them a while to get used to moving around because the moon has no gravity. They also collected about 48 pounds of moon rock, so that it could be studied by scientists back on earth.

REMEMBERING DETAILS

1) T
2) F The U.S. space program started with the Mercury program.
3) F Apollo 11 took off from Cape Kennedy, Florida. It landed in the Pacific Ocean.
4) F Two astronauts collected moon rock, which could then be studied by scientists back on earth.

INFERENCES

1) d
2) a

WORD POWER

1) c
2) a
3) c
4) b
5) b
6) a

CROSSWORD PUZZLE

ACROSS: 1) landing 4) astronauts 6) moon 7) power 9) descent 11) earliest 12) module 14) spacesuits 15) ladder

DOWN: 2) ignited 3) scientists 5) success 6) manned 8) orbits 10) cautiously 13) lunar

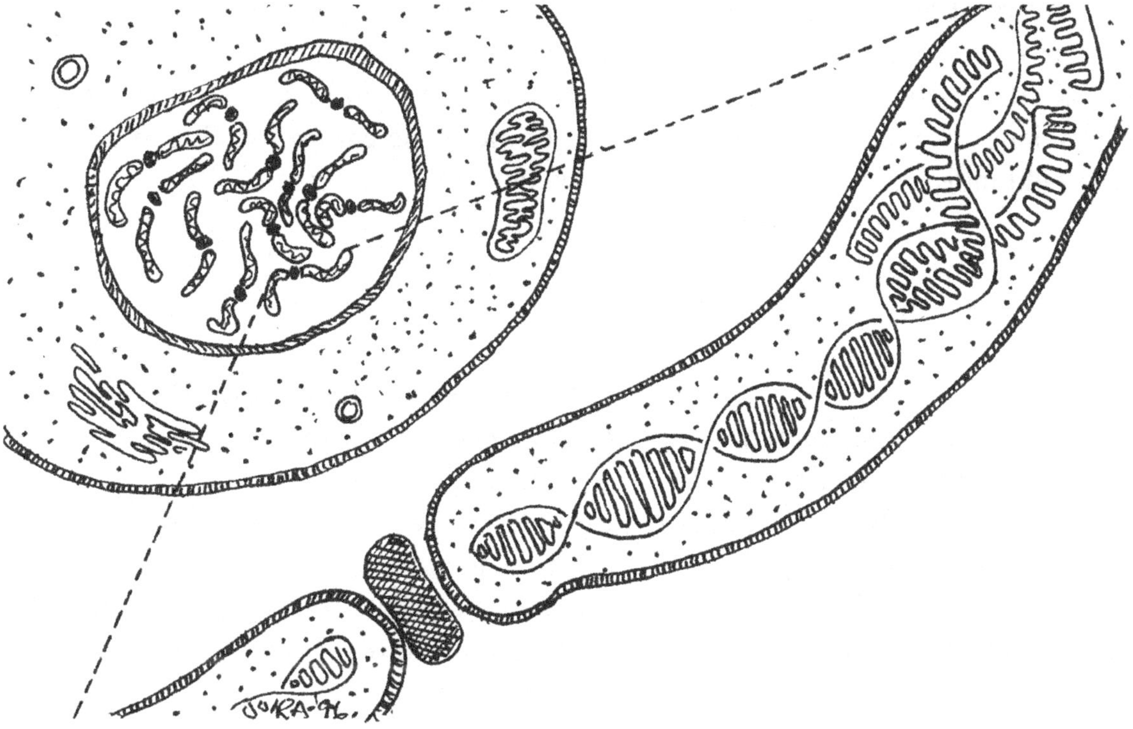

DNA

1 Every living thing in nature has the ability to reproduce, to make copies of itself. The blueprint, or plan followed by such reproduction is called heredity. And deoxyribonucleic acid (DNA) is the molecule that carries the genetic information needed for heredity.

2 For a long time scientists believed that the genes in our chromosomes contained not only the basic ability but also all the information needed for cells to reproduce. In 1951, Rosalind Franklin, a British biophysicist, discovered the molecular structure of DNA and its role as the key ingredient of the hereditary process.

3 DNA is made up of two matched strands that are coiled around each other and look like a twisting, spiral ladder called a double helix. These strands work together to make protein enzymes, which are the building blocks for creating cells.

One human cell consists of over 4
100,000 genes, and each of these genes represents a specific function or trait. When the cells begin to reproduce, or split, each new cell takes with it exactly the same information as the original cell had. For this to happen, the DNA strand at the heart of the cell must first separate, or "unzip," one half for the original cell, one for the new one. Then the new cell is able to grow and live on its own, having received a total genetic package of duplicate information.

5 The discovery of DNA and its role in heredity led to the creation of a new scientific field: genetic engineering. This new science figured out a way to remove or add individual parts of the DNA double helix or genetic code. This technique made it possible for the production of many medically useful substances.

6 The first genetically engineered product to be approved for human use was synthetic insulin. In 1980, the first diabetic patient in the world was injected with human-made insulin. Later another medicine, interferon, was successfully created. Interferon attacks viral infections and there is a possibility that it might be effective against some cancers.

7 Another field that has benefited from the discovery of DNA is police work and crime fighting. Since DNA is the basic genetic material found in all body cells, the DNA code for each individual is as unique as a fingerprint. In 1985, British geneticist Alec Jeffreys announced that chemical examination of blood, semen and other body fluids at a crime scene can be matched with a suspect's DNA and give positive identification. This method has been used with great success in rape cases and to a lesser degree in murder trials.

8 The understanding of how DNA works may one day make it possible for genetic scientists to eliminate inherited diseases at the prenatal stage of pregnancy. And the science of gene therapy is currently being used to help eliminate life-threatening diseases in children and adults.

9 But the most important future use of genetic engineering will probably be the successful identification and isolation of cancer-causing DNA. Medical advances in the recent past have given scientists hope for a new way of destroying the plague of these unwanted genes.

THE MAIN IDEA

Circle the letter of the sentence or phrase which best describes the main idea of the article about DNA. Be prepared to support your answer.

a) The structure of DNA.
b) Workings and applications of DNA.
c) The role of DNA in cell reproduction.
d) Medical and police use of knowledge about DNA.

UNDERSTANDING WHAT YOU READ

If you can, answer these questions from memory. If you cannot, look back at the article.

1) What does DNA do?

---

2) Why do new cells have exactly the same information as the cells from which they have divided?

---

3) What is genetic engineering?

---

4) What was the first genetically engineered product approved for human use?

REMEMBERING DETAILS

Write TRUE or FALSE under each statement. If the statement is false, write the statement correctly.

1) The British scientist, Rosalind Franklin, invented the field of genetic engineering.

---

2) DNA is made up of two exactly opposite strands in the form of a double helix.

---

3) Police have had the same success using DNA testing in rape and in murder cases.

---

4) Many people feel that the most important future application for genetic engineering will be in the fight against cancer.

INFERENCES

Based on the article, circle the letter of the best sentence completion.

1) The original discovery of DNA by Rosalind Franklin...

a) made all the later knowledge about DNA possible.
b) was a small thing compared with what was found out later.
c) was one of the most important applications of genetics ever.
d) had very little meaning for most scientists.

2) Genetic engineering and gene therapy seem...

a) to have reached the peak of their usefulness at this time.
b) to have brought us very few advantages so far.
c) to have been more important to scientists than to ordinary people.
d) to have shown the first of what will likely be many valuable applications.

INTERPRETATION

1) From human-made medicines to specially designed fruits and vegetables, genetic engineering is becoming more and more common. How do you feel about these developments? Will they help us, or will they perhaps get out of control?

2) Would you be interested in going into a career related to genetics? Why or why not?

3) Do you think every citizen should be required to give a sample of their DNA to the government for identification puposes and for the police to use in fighting crime? Should the government have the power to do this?

WORD POWER

Circle the letter of the word that means the same as the word on the left.

1) strands	a) threads	b) pieces	c) blocks
2) trait	a) problem	b) feature	c) part
3) split	a) divide	b) leave	c) die
4) duplicate	a) genetic	b) essential	c) copied
5) prenatal	a) midway	b) final	c) before birth
6) advances	a) experiments	b) breakthroughs	c) studies

CROSSWORD PUZZLE

ACROSS:

1) Rosalind Franklin discovered the ____________________ structure of DNA.

3) The strands work together to make protein ____________________.

5) ____________________ might be effective against some cancers.

8) One human ____________________ consists of over 100,000 genes.

10) The first genetically engineered product to be approved for humans was ____________________.

13) The blueprint, or plan followed by such reproduction is called ____________________.

DOWN:

2) A ____________________ examination of blood and other body fluids can be matched with DNA.

4) Each gene represents a specific ____________________ or trait.

6) The DNA code for each individual is as unique as a ____________________.

7) Everything in nature (any living thing) has the ability to ____________________.

9) Interferon attacks ____________________ infections.

11) DNA testing has been used with great success in ____________________ cases.

12) The science of ____________________ therapy is currently being used to help eliminate diseases.

14) ____________________ is made up of two matched strands that are coiled around each other.

ANSWER KEY

THE MAIN IDEA

b) Workings and applications of DNA.

UNDERSTANDING WHAT YOU READ

1) DNA carries the genetic information needed for heredity.
2) New cells have exactly the same information as the cells from which they have divided because each cell receives one half of the DNA strand, which splits or "unzips" lengthways into two equal parts.
3) Genetic engineering is a scientific field in which researchers use the ability to remove or add individual parts of the genetic code.
4) The first genetically engineered product approved for human use was synthetic insulin, the drug used to treat diabetes. The first patient was injected in 1980.

REMEMBERING DETAILS

1) F In 1951 the British scientist, Rosalind Franklin, was the first one to discover the importance of DNA.
2) F DNA is made up of two exactly identical strands in the form of a double helix.
3) F Police have had more success with DNA testing in rape cases than they have had in murder cases.
4) T

INFERENCES

1) a
2) d

WORD POWER

1) a
2) b
3) a
4) c
5) c
6) b

CROSSWORD PUZZLE

ACROSS: 1) molecular 3) enzymes 5) interferon 8) cell 10) insulin 13) heredity

DOWN: 2) chemical 4) function 6) fingerprint 7) reproduce 9) viral 11) rape 12) gene 14) DNA

Sports Medicine

1 Injuries can happen at any time, any place. When they do occur everyone likes to get the best treatment to help them heal quickly and properly. For athletes, the need to receive proper medical care is crucial in order for them to continue their sports careers. Athletes depend on the knowledge of doctors who are specially trained in sports medicine, which is a field of medicine that has grown rapidly since the 1950s.

2 Sports medicine is more than just the treatment of injuries. It is also concerned with the prevention of injuries, the maintaining of a proper diet, the creation of individual exercise programs for an athlete and the mental preparation of the athlete. In general, sports medicine is relevant to all aspects of monitoring athletes while they are in training. The field of sports medicine includes nutrition, surgery, physical therapy, research, and orthopedics, which is the correction or cure of diseases or deformities of bones, joints and muscles.

The doctors who specialize in sports medicine are called sports traumatologists. These experts specialize in the care of injuries to the musculoskeletal system. They do physical examinations, diagnose injuries, and refer patients to surgeons if necessary. 3

Although people have been interested in sports medicine for many years, it actually became more specialized after World War II, with 4 great developments in the 1960s and 1970s. The modern idea of complete care for the athlete emerged from the widespread surge in sports participation over this time period.

5 The field of sports medicine is very broad because there are so many types of sports injuries and because each individual athlete's body is different--their make-up, build, immune system, etc. Because of this, virtually every injury is treated in a different way.

6 As injuries continue to occur and the sports medicine field grows, recovery methods are becoming more advanced. Two of the more modern methods of treatment are the hyperbaric chamber and magnetic resonance imaging.

7 A hyperbaric chamber is a cylindrical steel tube into which a person can enter. Inside the chamber, the athlete is exposed to high levels of oxygen. This promotes oxygenation of the blood and speeds recovery time.

8 Use of magnetic resonance imaging (MRI) in medicine began in the early 1980s. MRI presents a hazard-free way to get images of thin slices of the body and a reliable method of detecting injuries. It is a superior imaging technique because it does not use radiation or need any special dyes. MRI uses magnets to concentrate and focus on small areas of the body, which produces detailed images. Besides being used to diagnose sports injuries, MRI is capable of producing high-contrast pictures of the brain, heart, liver, kidneys and can detect things such as tumors.

9 It is quite possible that with the increased interest in sports, the field of sports medicine will become even more specialized. Sports medicine continues to grow and take care of the needs of athletes from the professional playing in front of a stadium full of people to the person working out at their local health club.

THE MAIN IDEA

Circle the letter of the sentence or phrase which best describes the main idea of the article about sports medicine. Be prepared to support your answer.

a) Sports medicine is concerned with several aspects of an athlete's training program.
b) The advancements in sports medicine since the 1950s.
c) Sports medicine and what it does for athletes.
d) The excellent overall care provided to athletes by sports traumatologists.

UNDERSTANDING WHAT YOU READ

If you can, answer these questions from memory. If you cannot, look back at the article.

1) Why is sports medicine more than just the treatment of injuries?

___

2) Why is virtually every injury treated differently?

___

3) What is a hyperbaric chamber?

___

4) Why is MRI a superior imaging technique?

___

REMEMBERING DETAILS

Write TRUE or FALSE under each statement. If the statement is false, write the statement correctly.

1) Athletes depend on the knowledge of doctors who are specially trained in sports medicine.

___

2) Orthopedics is concerned with an athlete's diet.

___

3) An MRI only works after special dyes are injected into the patient.

___

4) Sports medicine is only for the most talented professional athletes.

___

INFERENCES

Based on the article, circle the letter of the best sentence completion.

1) If an athlete is not properly treated after an injury, then...

a) her ability to compete could suffer.
b) she will not pay the doctors.
c) an MRI would help.
d) she will definitely need surgery.

2) MRI can be used for...

a) diagnosing patients who are not athletes.
b) promoting oxygenation of the blood.
c) curing injuries.
d) getting rid of cancer.

INTERPRETATION

1) When an average athlete gets injured, he or she may have to wait an extended period of time before getting a needed MRI or other treatment. When a professional athlete gets injured he or she does not have to wait to see a doctor. Is this fair? Explain your answer.

2) Have you ever had a sports-related injury? What was it? Was it painful? What treatment did you receive? How long did it take to heal?

3) What sport do you think has the most potential for injury? List 10 sports in order, from most to least dangerous. Are the more dangerous sports more popular for spectators? Why or why not?

WORD POWER

Circle the letter of the word that means the same as the word on the left.

1) crucial	a) critical	b) passing	c) prompt
2) prevention	a) establishment	b) abundance	c) avoidance
3) deformities	a) abnormalities	b) changes	c) stresses
4) emerged	a) stepped back	b) leaned	c) came forth
5) broad	a) extensive	b) spread out	c) closed
6) reliable	a) inflatable	b) remarkable	c) dependable

CROSSWORD PUZZLE

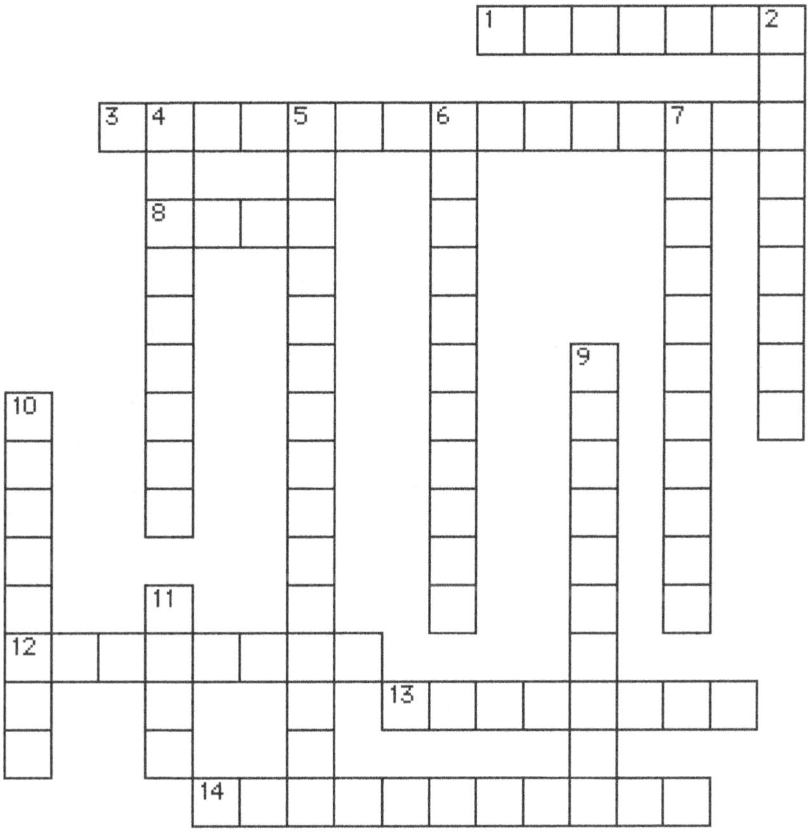

ACROSS:

1) Inside the ____________________, the athlete is exposed to high levels of oxygen.
3) The doctors who specialize in sports medicine are called sports ____________________.
8) MRI does not use radiation or need any special ____________________.
12) Sports medicine is more than just the treatment of ____________________.
13) MRI is a ____________________ imaging technique.
14) A hyperbaric chamber is a ____________________ steel tube into which a person can enter.

DOWN:

2) Use of magnetic ____________________ imaging in medicine began in the early 1980s.
4) MRI does not use ____________________.
5) Experts specialize in the care of injuries to the ____________________ system.
6) A hyperbaric chamber promotes ____________________ of the blood.
7) The field of sports medicine will become even more ____________________.
9) Two modern methods of treatment are the ____________________ chamber and MRI.
10) Athletes depend on the knowledge of doctors trained in sports ____________________.
11) Orthopedics is the correction or ____________________ of some diseases and deformities.

ANSWER KEY

THE MAIN IDEA

c) Sports medicine and what it does for athletes.

UNDERSTANDING WHAT YOU READ

1) Sports medicine is more than just the treatment of injuries because it is also concerned with the prevention of injuries, the maintaining of a proper diet, the creation of individual exercise programs for an athlete and the mental preparation of the athlete. In general, sports medicine is relevant to all aspects of monitoring athletes while they are in training. The field of sports medicine includes: nutrition, surgery, physical therapy, research and orthopedics.
2) Virtually every injury is treated in a different way because each individual athlete's body is different--their make-up, build, immune system, etc.
3) A hyperbaric chamber is a cylindrical steel tube into which a person can enter. Inside the chamber, the athlete is exposed to high levels of oxygen. This promotes oxygenation of the blood and speeds recovery time.
4) MRI is a superior imaging technique because it does not use radiation or need any special dyes.

REMEMBERING DETAILS

1) T
2) F Orthopedics is the correction or cure of diseases or deformities of bones, joints and muscles.
3) F An MRI does not need any special dyes.
4) F Sports medicine is for all athletes.

INFERENCES

1) a
2) a

WORD POWER

1) a
2) c
3) a
4) c
5) a
6) c

CROSSWORD PUZZLE

ACROSS: 1) chamber 3) traumatologists 8) dyes 12) injuries 13) superior 14) cylindrical

DOWN: 2) resonance 4) radiation 5) musculoskeletal 6) oxygenation 7) specialized 9) hyperbaric 10) medicine 11) cure

Hubble Space Telescope

1 What strange worlds lie beyond the moon, the sun and the stars? Are there planets in the solar system we have not yet found? Are there life forms living elsewhere deep in the galaxy? Is space an endless sea, or does it come to a sudden stop at some distant place in the heavens?

2 Like stars in the sky, there are millions of unanswered questions about the blackness of space that surrounds earth. Since the beginning of time, humans have tried to uncover the secrets that are hidden among the stars.

3 Many thousands of years spent gazing into the sky inspired humans to develop the science of astronomy, a field that involves the study of space. Many more hundreds of years passed before famed European astronomer Copernicus discovered that the planets revolve around the sun. But it was not until 1608, when a scientist named Galileo invented the very first telescope, that humans began enjoying their first real close up look at space.

Galileo's discovery opened up the skies for the first time, giving scientists a better view of the stars and planets in the solar system. Over the next three and a half centuries, humans explored space from the earth by looking through telescopes. That changed during the 20th century, when the invention of rockets and satellites allowed humans to travel into space for the first time. 4

5 By studying space, scientists learned that the ring of gases surrounding earth was stopping them for seeing a clear view of the heavens. This ring--called the earth's "atmosphere"--blocks many kinds of light from shining down to earth. In 1946, scientists put forth the idea of putting a telescope in space, high above the earth's atmosphere.

6 The Hubble Space Telescope is the end result of that idea. Launched in 1990 from Cape Canaveral in Florida, the Hubble Space Telescope travelled through the earth's atmosphere aboard the space shuttle. It was placed into orbit 325 miles (520 kilometers) above the earth's surface, giving science its first view of the universe from above the thick gases of the earth's atmosphere.

7 The Hubble Telescope took 12 years to develop, and cost over 2 billion dollars (U.S.) to make. It measures nearly 45 feet long (14 meters) and nearly 15 feet wide (5 meters). The telescope weighs over 24,000 pounds (11,000 kilograms), making it the largest civilian satellite ever released in space. It barely fit into the giant cargo bay of the space shuttle.

Despite being as large as a trailer 8
truck, the Hubble Space Telescope is able to point to any object in the sky with amazing accuracy. It will produce images of planets, stars and galaxies 20 times larger than the strongest telescope on earth.

The mirrors on the Hubble Telescope 9
are made of special glass that will not bend out of shape in space. This feature helps the telescope survive in conditions where there is no gravity. But despite its special design, even the smallest movement of the mirror's surface or sensors could obstruct its view. If that happens to the Hubble Space Telescope, humans will lose their eye in space!

THE MAIN IDEA

Circle the letter of the sentence or phrase which best describes the main idea of the article about the Hubble Space Telescope. Be prepared to support your answer.

a) Space has always fascinated humans.
b) Telescopes help people see distant planets.
c) The size of the Hubble Space Telescope.
d) The development of the Hubble Space Telescope.

UNDERSTANDING WHAT YOU READ

If you can, answer these questions from memory. If you cannot, look back at the article.

1) What did Copernicus do?

____

2) What happened in 1608?

____

3) When was the idea of putting a telescope into space first considered? When was the Hubble Space Telescope launched?

____

4) What is special about the mirrors on the Hubble Space Telescope?

____

REMEMBERING DETAILS

Write TRUE or FALSE under each statement. If the statement is false, write the statement correctly.

1) There are millions of unanswered questions about the blackness of space that surrounds earth.

____

2) Copernicus invented the very first telescope.

____

3) The Hubble Space Telescope fit easily into the space shuttle.

____

4) The Hubble Space Telescope will produce images of planets, stars and galaxies 100 times larger than the strongest telescope on earth.

____

INFERENCES

Based on the article, circle the letter of the best sentence completion.

1) The Hubble Space Telescope is the best way we have for viewing space because...

a) it is 20 times stronger than any telescope on earth.
b) it is out in space above the earth's atmosphere.
c) of its mirrors.
d) it cost 2 billion dollars (U.S.) to develop.

2) Scientists were able to put a telescope into space because...

a) Galileo invented the telescope.
b) the space shuttle is very large.
c) they could not see through the earth's atmosphere.
d) space travel made it possible.

INTERPRETATION

1) Do you think space exploration is important? Why or why not?

2) Two billion dollars could go a long way towards improving life for many people here on earth. Should this money have been spent on a telescope?

3) Before Copernicus discovered that the earth and other planets revolve around the sun, people believed that everything revolved around the earth. Can you think of any other scientific discoveries that have fundamentally changed our knowledge of our planet and our universe?

WORD POWER

Circle the letter of the word that means the same as the word on the left.

1) endless	a) closed	b) deep	c) infinite
2) gazing	a) frowning	b) peering	c) observing
3) famed	a) well-known	b) large	c) humble
4) barely	a) almost	b) just	c) nearly
5) accuracy	a) frankness	b) levitation	c) precision
6) obstruct	a) block	b) enhance	c) revolve

CROSSWORD PUZZLE

ACROSS:

5) The ____________________ blocks many kinds of light from shining down to earth.

9) The telescope is the largest ____________________ satellite ever released in space.

11) The smallest movement of the mirror's surface or ____________________ could obstruct its view.

13) Are there life forms living elsewhere deep in the ____________________?

14) The Hubble Space Telescope is the end result of an ____________________ that started in 1946.

DOWN:

1) There are millions of ____________________ questions about the blackness of space.

2) Galileo invented the very first ____________________.

3) Copernicus discovered that the planets ____________________ around the sun.

4) In 1946, scientists put forth the idea of ____________________ a telescope in space.

6) The ____________________ on the Hubble Telescope are made of special glass.

7) The invention of rockets and ____________________ allowed humans to travel into space.

8) Humans ____________________ space from the earth by looking through telescopes.

10) What ____________________ worlds lie beyond the moon, the sun and the stars?

12) It travelled through the earth's atmosphere ____________________ the space shuttle.

ANSWER KEY

THE MAIN IDEA

d) The development of the Hubble Space Telescope.

UNDERSTANDING WHAT YOU READ

1) Copernicus discovered that the planets revolve around the sun.
2) In 1608 a scientist named Galileo invented the very first telescope.
3) The idea of putting a telescope into space was first considered in 1946. The Hubble Space Telescope was launched in 1990.
4) The mirrors on the Hubble Space Telescope are made of special glass that will not bend out of shape in space. This feature helps the telescope survive in conditions where there is no gravity.

REMEMBERING DETAILS

1) T
2) F Galileo invented the very first telescope.
3) F The Hubble Space Telescope barely fit into the giant cargo bay of the space shuttle.
4) F The Hubble Space Telescope will produce images of planets, stars and galaxies 20 times larger than the strongest telescope on earth.

INFERENCES

1) b
2) d

WORD POWER

1) c
2) b
3) a
4) b
5) c
6) a

CROSSWORD PUZZLE

ACROSS: 5) atmosphere 9) civilian 11) sensors 13) galaxy 14) idea

DOWN: 1) unanswered 2) telescope 3) revolve 4) putting 6) mirrors 7) satellites 8) explored 10) strange 12) aboard

Genetic Engineering

1 Human beings have always attempted to improve on nature. For thousands of years, farmers have selectively bred the best cows, pigs or sheep from their herds, in order to improve the quality of their animals. Likewise, important crops like wheat, rice or corn have been improved by sowing seeds from the strongest or most productive plants. The world's food supply depends on such successful genetic experiments.

2 However, recent scientific advances have increased the speed and effectiveness of this process. In fact, modern genetic engineering works at the level not of entire plants or animals, but of individual cells. DNA, the genetic material at the heart of each cell, can be re-designed -- within a single species, or by moving DNA segments from one species into a second -- so as to produce a plant or animal with special characteristics.

In the mid-1990s experiments in 3
cloning large mammals, particularly sheep, attracted much attention. A clone is an exact copy of a living creature, produced artificially. People began to wonder how long it would be until even human clones were possible, and laws were passed to prohibit cloning experiments with humans. Still, at the cellular level cloning is the key to numerous benefits that we now take for granted, even if we do not always understand their origin.

Genetic engineers typically work to 4
alter the DNA of a plant or animal so

that a new genetic feature will always be reproduced exactly the same. This is molecular cloning. For instance, faster-growing food fish have been developed by introducing growth-regulating DNA from other species that grow particularly quickly. Also, the ability of certain fish to live in very cold conditions has been transferred to frost-sensitive plants; these crops can now be grown in areas where the temperature sometimes drops to levels that they normally could not tolerate.

5 Moreover, apart from simply increasing the volume of food, genetic engineering aims to reduce losses. Many food plants have been altered to escape harm by the herbicides that are used to kill weeds growing among them. And cereal crops, such as wheat or rice, are commonly engineered for reduced production of the enzymes that once caused them to spoil when ripe. Now, they can safely be stored for a longer time. Even the tomatoes that we buy from the supermarket may have been genetically engineered in this manner.

6 Medicine, as well, has benefitted from genetic engineering. Doctors and their patients rely on such important drugs as insulin, interferon, and human growth hormone. Once, these substances could only be collected very slowly from blood samples; now, they can be produced quickly and less expensively through the use of bacteria that have been genetically engineered with the relevant human DNA fragments. And other drugs are even produced in the milk of genetically altered cows or sheep.

7 Many interesting questions arise from this seemingly wonderful situation. Will the miracle of genetic engineering always be used positively, for the benefit of mankind? Could accidental consequences cause a genetic disaster, such as a new bacterium somehow resistant to all medicines? Do we in fact know enough about the possible results, good or bad? As time goes on, it appears that increasing scientific knowledge and public awareness will together guide the development of genetic engineering in the right direction.

THE MAIN IDEA

Circle the letter of the sentence or phrase which best describes the main idea of the article about genetic engineering. Be prepared to support your answer.

a) Genetic engineering can be extremely valuable if used appropriately.
b) Farmers have used a kind of genetic engineering for many years without realizing it.
c) Cloning is the most important form of genetic engineering.
d) It is necessary to improve scientific knowledge and public awareness before genetic engineering is developed any further.

UNDERSTANDING WHAT YOU READ

If you can, answer these questions from memory. If you cannot, look back at the article.

1) Traditionally, how have farmers tried to improve their animals and plants?

2) What is DNA?

3) What are two examples of how genetic engineering has been used to reduce losses in agriculture?

4) What medical drugs have been produced through genetic engineering?

REMEMBERING DETAILS

Write TRUE or FALSE under each statement. If the statement is false, write the statement correctly.

1) Human beings have always been content to leave nature alone.

2) Cereal crops, such as wheat or rice, are commonly engineered to have smaller crops.

3) Medicine has benefitted from genetic engineering.

4) Scientists have failed to produce drugs in the milk of genetically altered cows or sheep.

INFERENCES

Based on the article, circle the letter of the best sentence completion.

1) Genetic engineering is one example of...

a) cloning.
b) how humans have tried to improve on nature.
c) how to produce a plant or animal with special characteristics.
d) good ideas gone bad.

2) There is a fear that genetic engineering could...

a) disappear.
b) grow larger than the world's food supply.
c) be used for purposes other than for the benefit of mankind.
d) become illegal.

INTERPRETATION

1) Is genetic engineering going in the right direction? Discuss.

2) Do you think humans will be cloned in your lifetime? Would you allow yourself to be cloned?

3) Have humans done too much to alter nature? Discuss.

WORD POWER

Circle the letter of the word that means the same as the word on the left.

1) sowing	a) infusing	b) planting	c) detecting
2) segments	a) parts	b) puzzles	c) mobiles
3) key	a) argument	b) solution	c) charity
4) transferred	a) modified	b) exchanged	c) featured
5) manner	a) place	b) event	c) way
6) altered	a) changed	b) increased	c) fortified

CROSSWORD PUZZLE

ACROSS:

3) Will the miracle of genetic engineering always be used ____________________?

5) Scientific ____________________ have increased the speed and effectiveness of this process.

7) A clone is an exact copy of a living creature, produced ____________________.

9) ____________________, as well, has benefitted from genetic engineering.

10) Human beings have always attempted to ____________________ on nature.

11) They can safely be ____________________ for a longer time.

12) The temperature sometimes ____________________ to levels that they normally could not tolerate.

13) The world's food supply ____________________ on such successful genetic experiments.

DOWN:

1) People began to ____________________ how long it would be until human clones were possible.

2) Do we in fact know enough about the ____________________ results, good or bad?

4) Many interesting questions arise from this seemingly wonderful ____________________.

6) In the mid-1990s experiments in cloning large ____________________, particularly sheep, attracted much attention.

7) Many food plants have been ____________________ to escape harm by the herbicides that are used to kill weeds growing among them.

8) Once, these substances could only be ____________________ very slowly from blood samples.

ANSWER KEY

THE MAIN IDEA

a) Genetic engineering can be extremely valuable if used appropriately.

UNDERSTANDING WHAT YOU READ

1) By selectively breeding their best animals, and sowing seeds from their best plants.
2) The genetic material at the heart of each cell.
3) Food plants have been genetically engineered for resistance to herbicides used on weeds among them;some crops have been altered so that they produce less of the enzymes that cause them to spoil when ripe.
4) Insulin, interferon, and human growth hormone.

REMEMBERING DETAILS

1) F Human beings have always attempted to improve on nature.
2) F Cereal crops, such as wheat or rice, are commonly engineered for reduced production of the enzymes that once caused them to spoil when ripe.
3) T
4) F Drugs are produced in the milk of genetically altered cows or sheep.

INFERENCES

1) b
2) c

WORD POWER

1) b
2) a
3) b
4) b
5) c
6) a

CROSSWORD PUZZLE

ACROSS: 3) positively 5) advances 7) artificially 9) medicine 10) improve 11) stored 12) drops 13) depends

DOWN: 1) wonder 2) possible 4) situation 6) mammals 7) altered 8) collected

Endangered Species

1 Imagine a single goldfish swimming in a pond of fresh water. If its food and water supply remained the same, this fish could live a long happy life. But what if conditions in the pond were to change?

2 What if gasoline were poured into the water, poisoning the fish? What if the pond were drained away, leaving the fish without water? What if hundreds of other goldfish were crowded into the pond, or a hungry trout looking to have fresh fish for dinner were put into the pond? In each case, the goldfish would probably disappear from the pond forever.

3 When something changes the natural conditions in which groups of plants or animals are used to living, it can lower the chances these groups--or "species" --will survive. This destruction of wild lands and natural habitats is a major reason why some plants and animals become "endangered species."

4 An endangered species is a living thing--plant, animal or even insect-- that is close to becoming "extinct." This means their entire species is close to disappearing from the planet forever. Dinosaurs are the most common example of an extinct species. These giant beasts suddenly disappeared from earth over 65 million years ago and can no longer be found living anywhere on the planet.

5 A more recent example would be the carrier pigeon. Once found by the millions in North America, the number

of carrier pigeons began dropping in the 19th century. The major cause was over-hunting by humans and the destruction of the forests where they once lived. Although efforts were made to save the bird, the last carrier pigeon died in 1914.

6 By the year 2000 there were close to one million fewer species on earth than there were at the beginning of that century. In the 1980s, the world was losing one species a day. In the 1990s, the number increased to one species every hour! Over 1,000 of the 9,000 known species of birds are now on the endangered species list. So are 25,000 of the 250,000 known types of plants. Humpback whales, African elephants, Siberian tigers, lions, zebras, cougars, eagles, falcons, giraffes, crocodiles, chinook salmon and wild orchids are just a few of the plants, animals, birds and fish that are currently facing extinction.

There are many things that can put a 7
species in danger. There are natural causes, such as earthquakes, floods, droughts, volcanoes, fire and hunting by other animals. However, humans are by far the greatest cause of destruction in the plant and animal kingdom.

While humans have always relied on 8
hunting and gathering for food, over-hunting and killing animals for sport has brought many species close to extinction. Clearing forests for farmland, for firewood and lumber has destroyed millions of acres of natural habitat. This is especially true in the tropical rain forests, where nearly half the species on earth are found. Even in water, over-fishing has decreased the number of fish swimming in the lakes and oceans of the earth. If this destruction continues, there are many people who predict humans themselves will be the next endangered species!

THE MAIN IDEA

Circle the letter of the sentence or phrase which best describes the main idea of the article about endangered species. Be prepared to support your answer.

a) Humans are quickly becoming endangered.
b) The problem of species becoming extinct.
c) Many things cause a species to become endangered.
d) Species have been disappearing for millions of years.

UNDERSTANDING WHAT YOU READ

If you can, answer these questions from memory. If you cannot, look back at the article.

1) Name something which can lower the chances that a species survives.

____

2) What is an endangered species?

____

3) Why are scientist concerned about disappearing species when this has been happening for millions of years?

____

4) What are some natural causes that can put a species in danger of extinction?

____

REMEMBERING DETAILS

Write TRUE or FALSE under each statement. If the statement is false, write the statement correctly.

1) Carrier pigeons became extinct 65 million years ago.

____

2) By the year 2000, there were close to one thousand fewer species on earth than there were at the beginning of that century.

____

3) Natural causes are the greatest cause of destruction in the plant and animal kingdoms.

____

4) There are many people who predict mankind itself will be the next endangered species.

____

INFERENCES

Based on the article, circle the letter of the best sentence completion.

1) The speed at which species are becoming extinct is a concern because...

a) most species of birds are now on the endangered list.
b) it will not go away unless we do something about it.
c) rain forests are being cleared so quickly.
d) species are disappearing at a faster rate than ever before.

2) The problem of endangered species...

a) will be solved by plants and animals.
b) is the fault of the dinosaurs.
c) is mostly a human-made problem.
d) has slowed in recent years.

INTERPRETATION

1) What can an individual person do to help the cause of preventing species from becoming extinct?

2) How does recycling aid in the preservation of plant and animal life?

3) Do you think humans are truly in danger of becoming extinct? Why or why not?

WORD POWER

Circle the letter of the word that means the same as the word on the left.

1) dropping	a) falling	b) skyrocketing	c) jumping
2) predict	a) worry	b) anticipate	c) know
3) facing	a) encountering	b) preventing	c) hurrying
4) droughts	a) wet seasons	b) dry spells	c) cold seasons
5) relied on	a) counted on	b) estimated	c) pretended
6) clearing	a) mowing	b) cutting down	c) addressing

CROSSWORD PUZZLE

ACROSS:

1) Many people predict humans themselves will be the next ____________________ species!

5) In the 1980s, the world was ____________________ one species a day.

6) Nearly half the species on earth are found in ____________________ rain forests.

9) There are many things that can put a ____________________ in danger.

11) ____________________ disappeared from earth over 65 million years ago.

12) An endangered species is a ____________________ thing that is close to becoming extinct.

13) 25,000 of the 250,000 known types of ____________________ are on the endangered species list.

14) Imagine a single goldfish ____________________ in a pond of fresh water.

DOWN:

2) Humans are by far the greatest cause of ____________________ in nature.

3) Even in water, over-fishing has ____________________ the number of fish.

4) Carrier pigeons were once found by the ____________________ in North America.

7) What if gasoline was poured into the water, ____________________ the fish?

8) What if hundreds of other goldfish were ____________________ into the pond?

10) The destruction of wild lands and natural ____________________ is a major problem.

ANSWER KEY

THE MAIN IDEA

b) The problem of species becoming extinct.

UNDERSTANDING WHAT YOU READ

1) When something changes the natural conditions in which groups of plants or animals are used to living, it can lower the chances these species will survive.
2) An endangered species is a living thing--plant, animal or even insect--that is close to becoming "extinct." This means their entire species is close to disappearing from the planet forever.
3) Scientists are concerned about extinctions because even though this has been happening for millions of years, in the recent past the rate of species going extinct has speeded up considerably.
4) Some natural causes that can put a species in danger of extinction include earthquakes, floods, droughts, volcanoes, fire and hunting by other animals.

REMEMBERING DETAILS

1) F Carrier pigeons became extinct in 1914. Dinosaurs became extinct 65 million years ago.
2) F By the year 2000 there were close to one million fewer species on earth than there were at the beginning of that century.
3) F Humans are by far the greatest cause of destruction in the plant and animal kingdoms.
4) T

INFERENCES

1) d
2) c

WORD POWER

1) a
2) b
3) a
4) b
5) a
6) b

CROSSWORD PUZZLE

ACROSS: 1) endangered 5) losing 6) tropical 9) species 11) dinosaurs 12) living 13) plants 14) swimming

DOWN: 2) destruction 3) decreased 4) millions 7) poisoning 8) crowded 10) habitats

H[illegible]

1 If you are tired of watering your plants every day, you might want to try growing them right in the water. And in fact this is not a new idea: people in ancient Egypt were growing plants without soil more than 2,000 years ago. Moreover, the floating gardens of the Aztecs in early Mexico are another ancient example of gardening without soil. Nowadays this technique is widely used. It is called hydroponic gardening.

2 Literally, the term hydroponics means cultivating in water rather than in soil, and in fact normal soil is not used in this process. But plants are plants, and so the concerns of the hydroponic gardener are really much the same as those of any other farmer. Even when they do not grow in soil, all plants need food, stability, oxygen, and protection from disease.

The main nutrients required by plants 3 are nitrogen, potassium, and phosphorus. Also, small amounts of other elements are needed. In the ancient hydroponic gardens of Egypt and Mexico, the growers had to rely on minerals that occurred naturally in the water. In modern hydroponic applications, workers carefully add these important minerals to pure water; this way, the balance of waterborne nutrients can be controlled, depending on the kind of plants being grown and on the speed of growth of the crop.

Plants usually get stability from the 4 soil around their roots. Obviously, water cannot provide this. Plants farmed hydroponically are supported either by growing through a mesh over the water, or by sinking their roots into sand or pea-sized gravel soaked in water. Either way, the plants stay upright, with their leaves exposed to the sun. They need no soil because, in this system, they do not derive nutrients from the medium that gives them stability.

5 Oxygen is another essential requirement. The roots as well as the leaves of all plants must be able to breathe. If insufficient oxygen reaches its roots, a plant will die. To fulfill this need, hydroponic farmers usually keep the nutrient water flowing all the time. As the water circulates, it is aerated by splashing from one container to another: the splashing process increases the amount of absorbed air in the water.

6 Finally, plants need a healthy environment. Especially when plants of the same species are grown close together, there is a real danger that diseases may spread quickly. A very clean growing medium will reduce this risk. Soil, the natural planting medium, may contain many harmful diseases or pests, but hydroponic gardening does not use soil. Starting with pure water and clean sand or gravel, the hydroponic gardener avoids many uncertainties.

7 In these ways, hydroponics can overcome important agricultural challenges. For instance, hydroponic farming is able to produce worthwhile crops in regions where the soil is too poor for ordinary gardens, or where the supply of water is too scarce for normal irrigation. Also, even in fertile areas, hydroponic greenhouses make it possible to cultivate small areas extremely densely: this can be very profitable during winter in cold climates, such as Canada or the Northern United States. Such considerations may not have mattered to the ancient Egyptians or Aztecs, but they are the basis of a thriving hydroponics industry today.

THE MAIN IDEA

Circle the letter of the sentence or phrase which best describes the main idea of the article about hydroponics. Be prepared to support your answer.

a) Hydroponic gardening is ideal for locations where water is too scarce for normal irrigation.
b) Hydroponic gardening solved the problem of plant stability.
c) Hydroponic gardening makes it possible for plants to be grown inside in countries with a cold winter.
d) Hydroponic gardening is not only a very old technique but also a very modern one with great future potential.

UNDERSTANDING WHAT YOU READ

If you can, answer these questions from memory. If you cannot, look back at the article.

1) When did people first start growing plants without soil?

____

2) What four things do all plants need?

____

3) Why can hydroponically grown plants do without soil?

____

4) How do hydroponic farmers assure that oxygen reaches the roots of their plants?

____

REMEMBERING DETAILS

Write TRUE or FALSE under each statement. If the statement is false, write the statement correctly.

1) Plants that are grown without soil do not need food.

____

2) The roots as well as the leaves of all plants must be able to breathe.

____

3) When plants are grown close together there is less of a chance for disease.

____

4) Soil is a much cleaner planting medium than is hydroponic gardening.

____

INFERENCES

Based on the article, circle the letter of the best sentence completion.

1) Hydroponic gardening...

a) is still used by the Aztecs and the Egyptians.
b) takes less room than conventional soil gardening.
c) results in many unstable plants.
d) is popular in Canada only during the winter.

2) Plants grown hydroponically...

a) usually have a much bigger harvest.
b) are more expensive than plants grown in soil.
c) are also called "organic" fruits and vegetables.
d) are really no different than plants grown in soil.

INTERPRETATION

1) How do you think hydroponic gardening could be used to help people suffering due to hunger? Discuss.

2) Do you enjoy gardening? Why? Why not? What do you like to grow? What would you grow if you could have your dream garden? List ten things you would grow.

3) Do you think a day will come when all the plants from which humans derive food will be grown indoors hydroponically? Describe such a world.

WORD POWER

Circle the letter of the word that means the same as the word on the left.

1) stability	a) solidness	b) awkwardness	c) variation
2) balance	a) greed	b) levels	c) containers
3) supported	a) picked	b) leaned	c) braced
4) derive	a) get	b) expel	c) revolve
5) absorbed	a) taken in	b) implied	c) breathed
6) fertile	a) moderate	b) rich	c) barren

CROSSWORD PUZZLE

ACROSS:

3) The ____________________ gardens are an example of gardening without soil.

4) Soil may contain many ____________________ diseases or pests.

7) Literally, the term hydroponics means ____________________ in water rather than in soil.

12) All plants must be able to ____________________ .

13) People in ____________________ Egypt were growing plants without soil 2,000 years ago.

DOWN:

1) Hydroponic greenhouses make it possible to cultivate small areas ____________________ .

2) Plants usually get ____________________ from the soil around their roots.

4) Plants need a ____________________ environment.

5) The concerns of the ____________________ gardener are really much the same as those of any other farmer.

6) The growers had to rely on ____________________ that occurred naturally in the water.

8) The plants stay ____________________ , with their leaves exposed to the sun.

9) Hydroponic farmers usually keep the ____________________ water flowing all the time.

10) A very clean growing medium will ____________________ this risk

11) Hydroponics can ____________________ important agricultural challenges.

ANSWER KEY

THE MAIN IDEA

d) Hydroponic gardening is not only a very old technique but also a very modern one with great future potential.

UNDERSTANDING WHAT YOU READ

1) Over 2,000 years ago.
2) Food, stability, oxygen and protection from disease.
3) They do not derive food from the medium that gives them stability.
4) They keep the nutrient water constantly flowing, and splashing from one container to another.

REMEMBERING DETAILS

1) F Even when they do not grow in soil, all plants need food, stability, oxygen, and protection from disease.
2) T
3) F Especially when plants of the same species are grown close together, there is a real danger that diseases may spread quickly.
4) F Soil, the natural planting medium, may contain many harmful diseases or pests, but hydroponic gardening does not use soil. Starting with pure water and clean sand or gravel, the hydroponic gardener avoids many uncertainties.

INFERENCES

1) b
2) d

WORD POWER

1) a
2) b
3) c
4) a
5) a
6) b

CROSSWORD PUZZLE

ACROSS: 3) floating 4) harmful 7) cultivating 12) breathe 13) ancient

DOWN: 1) densely 2) stability 4) healthy 5) hydroponics 6) minerals 8) upright 9) nutrient 10) reduce 11) overcome

Satellites

1 On October 4, 1957 the Space Age began. That was the day that the former Soviet Union launched the first human-made satellite, Sputnik I. Today estimates range between 8 and 25 thousand objects orbiting the earth.

2 A satellite is an object that revolves around a larger planet or star. There are two kinds of satellites: natural and artificial. Natural satellites include the moons around the planets, including ours. Of course, the earth is a natural satellite of the sun. Artificial satellites include all the objects humans have blasted into space that continue to revolve around the earth or any other planet. Famous artificial satellites include the Hubble Space Telescope and the space stations, Skylab, Salyut and Mir, where astronauts have lived and conducted experiments while orbiting our planet.

When the first satellites were 3
launched in the late 1950s, scientists were mostly concerned with proving that satellites were more than just science fiction. The cargo of Sputnik I was a thermometer and a radio transmitter. That transmitter was used to beam a radio signal back to the earth. Today some satellites can carry over 120,000 different messages at once, and there is a network of communication satellites that is shared by more than 100 nations around the globe.

Launching satellites is usually done 4
with multistage rockets. These rockets must reach speeds of over 5 miles per

second (over 8 kilometers per second) to propel the satellite to a height of at least 100 miles (160 kms) above the earth. At that distance the satellite will no longer be affected by the earth's outer atmosphere, and will begin revolving around the planet in a nearly circular path.

5 A satellite that is lifted 22,300 miles (35,700 kms) into space will take exactly 24 hours to revolve around the earth. A satellite at that distance that has its orbit over the equator and rotates in the same direction as the earth will remain at the same spot over the earth's surface. Such a satellite is said to be in a geostationary orbit.

6 At times the position of a satellite must be adjusted to compensate for movements caused by forces at work in outer space. A satellite can be affected by such things as the sun, the moon, the planets, and any space debris that might come into contact with it. Small rockets built into the satellite can be fired to move the satellite back to the exact spot where it is supposed to be located.

There are five principal types of 7
artificial satellites: communications, scientific, military, weather, and navigation. Communications satellites carry telephone, television and radio signals. Scientific satellites are used to study outer space, such as the Hubble Space Telescope. Military satellites, which are sometimes called spy satellites, are used to spy on a country's enemies. Weather satellites transmit data that provides meteorologists with detailed weather information. And navigation satellites are used by airplanes and ships to pinpoint their positions with accuracy.

The satellite industry is still in its 8
infancy. Who knows what future satellites will be able to do?

THE MAIN IDEA

Circle the letter of the sentence or phrase which best describes the main idea of the article about satellites. Be prepared to support your answer.

a) Artificial versus natural satellites.
b) The five principal types of satellites.
c) The global communication satellite system.
d) An introduction to satellites.

UNDERSTANDING WHAT YOU READ

If you can, answer these questions from memory. If you cannot, look back at the article.

1) When did the Space Age begin?

____

2) What is the difference between a natural and an artificial satellite?

____

3) How high must a satellite be propelled into space so as to not be affected by the earth's outer atmosphere?

____

4) What are the five principal types of artificial satellites?

____

REMEMBERING DETAILS

Write TRUE or FALSE under each statement. If the statement is false, write the statement correctly.

1) The United States launched the first human-made satellite.

____

2) The cargo of Sputnik was a thermometer and a radio transmitter.

____

3) Once in orbit, satellites remain in exactly the same position.

____

4) Military satellites are used by airplanes and ships to pinpoint their positions with accuracy.

____

INFERENCES

Based on the article, circle the letter of the best sentence completion.

1) It appears that in our modern world everyone will...

a) have to know a lot about satellites.
b) use many services provided through satellites.
c) be in danger because of satellites.
d) see more and more satellites up in the sky.

2) When the writer says that "...the satellite industry is still in its infancy," it means that...

a) the Space Age began in 1957.
b) satellites are getting larger and larger.
c) satellites will probably be able to do a lot more in the future.
d) children easily understand how satellites work.

INTERPRETATION

1) Some people are beginning to worry that, with all the satellites and other "space junk" in orbit around the earth, we are actually polluting outer space! Do you think this is a problem? Do we need rules?

2) Think about your activities over the past week: what different practical or recreational uses did you make of satellite-related technology?

3) The more we rely on satellites for communications, the less we know about how to use beautiful old fashioned means of communication such as letters. Are we losing something?

WORD POWER

Circle the letter of the word that means the same as the word on the left.

1) artificial	a) untrue	b) manufactured	c) chemical
2) network	a) system	b) schedule	c) grade
3) compensate	a) make up	b) renew	c) measure
4) debris	a) ships	b) gravity	c) junk
5) transmit	a) receive	b) interpret	c) send
6) pinpoint	a) locate exactly	b) adjust	c) surmount

CROSSWORD PUZZLE

ACROSS:

1) ____________________ satellites are used by airplanes and ships to pinpoint their positions.

5) There are five ____________________ types of artificial satellites.

7) Launching satellites is usually done with ____________________ rockets.

8) Who knows what ____________________ satellites will be able to do?

10) A satellite is an object that ____________________ around a larger planet or star.

13) At times the position of a satellite must be ____________________.

DOWN:

2) The satellite will no longer be ____________________ by the earth's outer atmosphere.

3) The cargo of Sputnik I was a thermometer and a radio ____________________.

4) There are two kinds of satellites: ____________________ and artificial.

6) On October 4, 1957 the Space Age ____________________.

8) ____________________ artificial satellites include the Hubble Telescope and the space stations.

9) Scientists were mostly concerned with ____________________ that satellites could work.

11) Scientific satellites are used to study ____________________ space.

12) Military satellites are used to ____________________ on a country's enemies.

ANSWER KEY

THE MAIN IDEA

d) An introduction to satellites.

UNDERSTANDING WHAT YOU READ

1) The Space Age began on October 4, 1957. That was the day the former Soviet Union launched the first human-made satellite, Sputnik I.
2) The difference between a natural satellite and an artificial satellite is that a natural satellite is a satellite that originated in space, whereas an artificial satellite originated on earth and was put into orbit by humans.
3) A satellite must be propelled at least 100 miles (160 kms) into space so as to not be affected by the earth's outer atmosphere.
4) The five principal types of artificial satellites are communications satellites, scientific satellites, military satellites, weather satellites, and navigation satellites.

REMEMBERING DETAILS

1) F The former Soviet Union launched the first human-made satellite.
2) T
3) F Once in orbit, satellites do not always remain in the same position. Satellite movements can be caused by forces at work in outer space.
4) F Military satellites, which are sometimes called spy satellites, are used to spy on a country's enemies. Navigation satellites are used by airplanes and ships to pinpoint their positions with accuracy.

INFERENCES

1) b
2) c

WORD POWER

1) b
2) a
3) a
4) c
5) c
6) a

CROSSWORD PUZZLE

ACROSS: 1) navigation 5) principal 7) multistage 8) future 10) revolves 13) adjusted

DOWN: 2) affected 3) transmitter 4) natural 6) began 8) famous 9) proving 11) outer 12) spy

Rain Forests

1 In recent times human activity has had a greater impact on the world's rain forests than ever before. Slash-and-burn agriculture, where trees are chopped down and everything else is burned to prepare the land for planting crops, as well as logging and mining are responsible for a great deal of the damage. When rain forests are cleared, the ecosystem is thrown out of balance, species of plants and animals become endangered, and the planet loses more of its natural filters--the trees--that prevent global warming.

2 There are basically two kinds of rain forest: temperate and tropical. Temperate means that the climate gets neither very hot nor very cold. Large temperate rain forests are located in Australia, New Zealand, Chile, the United States, and Canada. The largest tropical rain forests in the world are the famous forest around the Amazon river in South America, and the forest around the Congo river in Africa. In tropical rain forests the temperature is constantly hot, and it rains often.

The climate in a tropical rain forest 3
makes it a place with an abundance of both plant and animal life. Even today scientists are discovering never-before-known species in the Amazonian rain forest; it is one of the last places on the planet not to be understood completely. In general, tropical rain forests are very humid. Temperatures average about 80 degrees Fahrenheit (26 degrees Celsius). Couple this with the more than 60 inches (150 centimeters) of rain that a tropical rain forest

receives each year, and it is easy to understand where the humidity comes from.

4 The trees in a tropical rain forest grow in layers. The top layer is made up of only the tallest trees in the forest. These can grow as tall as 200 feet (61 meters). Below the taller trees, on the next layer, is a vast population of shorter trees, between 60 and 100 feet tall (18 to 30 meters), which form a shell, or canopy. In many places there is a second canopy below this upper one. Because of the trees and other vegetation, the canopies are so thick at some places that no sunlight reaches the forest floor. Many species of vines live in almost total darkness beneath the canopies, making the rain forest a real jungle!

5 When acres of rain forest are clear-cut for agriculture, the soil to be farmed only contains enough nutrients to support two or three plantings, after which the land will be abandoned, and unless reforestation takes place, vines and coarse grasses will overtake the plot and part of the forest will disappear forever. In recent times, human activity has been responsible for destroying between 20 and 50 million acres of rain forest each year.

6 Scientists figure that over 40,000 species of plants grow in the Amazonian rain forest. Studies estimate that in tropical rain forests around the world a species goes extinct once every six hours. Think what would happen if one of the species which went extinct could have been made into a medicine to cure cancer or could have held some other benefit for humankind?

7 Today rain forests cover less than six percent of the earth's total land surface. Remember, through photosynthesis green plants supply humans with oxygen to breathe. The rain forests are the world's biggest provider of clean air. Can we survive without them?

THE MAIN IDEA

Circle the letter of the sentence or phrase which best describes the main idea of the article about rain forests. Be prepared to support your answer.

a) Human activity in the rain forests of the world.
b) Plant life in rain forests.
c) Information about rain forests.
d) The total destruction of the world's rain forests.

UNDERSTANDING WHAT YOU READ

If you can, answer these questions from memory. If you cannot, look back at the article.

1) What happens when rain forests are cleared?

____

2) How many kinds of rain forest are there? What are they called?

____

3) What grows in almost total darkness beneath the forest canopy?

____

4) How often does a species go extinct in tropical rain forests around the world?

____

REMEMBERING DETAILS

Write TRUE or FALSE under each statement. If the statement is false, write the statement correctly.

1) In temperate rain forests the temperature is constantly hot, and it rains often.

____

2) Tropical rain forests are very humid.

____

3) Scientists estimate that 40,000 species of plants have already been made extinct in the Amazonian rain forest.

____

4) Green plants supply humans with oxygen through photosynthesis.

____

INFERENCES

Based on the article, circle the letter of the best sentence completion.

1) When it comes to using rain forest land for agriculture, the author seems to suggest that...

a) the canopies are so thick that the trees must be cut down.
b) slash-and-burn is the best method.
c) the forest grows back automatically.
d) the land is not valuable enough as farmland to warrant losing the rain forest.

2) When the author calls the rain forest a "real jungle" it means that the rain forest is...

a) totally dark.
b) a great place for animals to live.
c) a tangle of plant life.
d) a scary place.

INTERPRETATION

1) Are there rain forests in your country? How do they benefit people?

2) Can we survive without the rain forests? How can we prevent their destruction?

3) How have people living near and in the Amazonian rain forest been affected by human activity?

WORD POWER

Circle the letter of the word that means the same as the word on the left.

1) impact	a) neglect	b) effect	c) misfortune
2) constantly	a) always	b) rarely	c) increasingly
3) abundance	a) uniqueness	b) speciality	c) wealth
4) vegetation	a) creatures	b) plants	c) vegetables
5) nutrients	a) filters	b) organs	c) minerals
6) abandoned	a) left behind	b) boarded up	c) torn apart

CROSSWORD PUZZLE

ACROSS:

1) Over 40,000 species of ____________________ grow in the Amazonian rain forest.
3) In many places there is a second ____________________ below the upper one.
4) The canopies are so thick that no ____________________ reaches the forest floor.
6) Acres of rain forest is clear-cut for ____________________.
12) Rain forests cover less than six ____________________ of the earth's land surface.
13) Can we ____________________ without the rain forests?

DOWN:

1) The rain forests are the world's biggest ____________________ of clean air.
2) When rain forests are cleared the ecosystem is thrown out of ____________________.
5) Tropical rain forests are very ____________________.
7) Vines and coarse ____________________ will overtake the plot.
8) It is one of the last places on the planet not to be ____________________ completely.
9) There are basically two kinds of rain forest: ____________________ and tropical.
10) The trees in a tropical rain forest grow in ____________________.
11) Studies estimate that a species goes ____________________ once every six hours.

ANSWER KEY

THE MAIN IDEA

c) Information about rain forests.

UNDERSTANDING WHAT YOU READ

1) When rain forests are cleared, the ecosystem is thrown out of balance, species of plants and animals become endangered, and the planet loses more of its natural filters--the trees--that prevent global warming.
2) There are two kinds of rain forest. They are called temperate and tropical.
3) Vines grow in almost total darkness beneath the canopy.
4) Studies estimate that in tropical rain forests around the world a species goes extinct once every six hours.

REMEMBERING DETAILS

1) F In temperate rain forests the climate gets neither very hot nor very cold. In tropical rain forests the temperature is constantly hot, and it rains often.
2) T
3) F Scientists figure that over 40,000 species of plants grow in the Amazonian rain forest.
4) F Through photosynthesis green plants supply humans with oxygen to breath.

INFERENCES

1) d
2) c

WORD POWER

1) b
2) a
3) c
4) b
5) c
6) a

CROSSWORD PUZZLE

ACROSS: 1) plants 3) canopy 4) sunlight 6) agriculture 12) percent 13) survive

DOWN: 1) provider 2) balance 5) humid 7) grasses 8) understood 9) temperate 10) layers 11) extinct

Made in the USA
Monee, IL
08 November 2026

68091709R00090